MW01489651

Quantitative Chemical Analysis

Thomas Edward Thorpe

BIBLIOBAZAAR

QUANTITATIVE

CHEMICAL ANALYSIS.

BY

T. E. THORPE, Ph.D., B.Sc.(Vict.), F.R.S.

PROFESSOR OF CHEMISTRY IN THE NORMAL SCHOOLS OF SCIENCE, SCIENCE AND
ART DEPARTMENT, SOUTH KENSINGTON, LONDON.

NINTH EDITION.

JOHN WILEY & SONS
53 EAST TENTH STREET,
NEW YORK.
1891.

PREFACE

THE FIFTH EDITION.

THE PRESENT EDITION of this work will be found to include a considerable amount of new matter. Many valuable hints and suggestions have been received from teachers and others both in this country and in America. Professor Frankland has kindly looked over the section on Water Analysis; Professor Dittmar has furnished an account of his method for the valuation of chrome ore; and Mr. Watson Smith has sent the results of his experience of the work in the laboratory of the Owens College, Manchester. To these and to other gentlemen who have furnished him with additions and corrections, the Author begs to tender his grateful acknowledgments. He also desires to express his indebtedness to Mr. C. H. Bothamley, Assistant-Lecturer on Chemistry in the Yorkshire College, for aid in the revision of the book.

YORKSHIRE COLLEGE, LEEDS:

PREFACE.

THE AIM of this book is to teach the principles of Quantitative Chemical Analysis by the aid of examples chosen, partly on account of their practical utility, and partly as affording illustration of the more important quantitative separations.

It is divided into five distinct parts. The first part gives a description of the balance, of the mechanical principles involved in its construction, and of the manner of using it. It also contains an account of the operations generally or most frequently occurring in Quantitative Analysis ; such as the process of filtration, the incineration of filters, and so forth.

The second part consists of a graduated series of examples in simple gravimetric analysis, commencing with the analysis of copper sulphate, and ending with the estimation of arsenic, antimony and tin.

The third part treats of volumetric analysis. The more important volumetric processes are fully described, and their application is illustrated by examples of scientific as well as of technical interest.

The fourth part contains an account of the methods, gravimetric and volumetric, employed in the analysis or valuation of ores, minerals, and of the more important industrial products, such as copper and lead ores, iron and manganese ores, limestone, cast and wrought iron, soda-ash, bleaching powder, &c. Considerable space has been allotted to the important subject of water-analysis.

The fifth part treats of the general processes of organic analysis.

The Author's thanks are due to his assistant, Mr. Dugald Clerk, for the attention he has bestowed on the drawings for the woodcuts. The illustrations in the section on 'Water-analysis' are taken, with Mr. Sutton's kind permission, from his work on 'Volumetric Analysis.' In the account of Frankland and Armstrong's method of determining the amount of organic carbon and nitrogen in water, it will be seen how much the Author is indebted to Mr. Wm. Thorp's excellent description of the process in that manual. Mr. Crookes has also kindly allowed the use of the figures in illustration of Luckow's process for assaying copper-ores.

CONTENTS.

---·◊·---

PART I.

---— — —. — - —

PART II.

Contents.

PART III.

SIMPLE VOLUMETRIC ANALYSIS OF SOLIDS AND LIQUIDS.

ALKALIMETRY.

PART IV.

*GENERAL ANALYSIS, INVOLVING GRAVIMETRIC AND
VOLUMETRIC PROCESSES.*

PART V.

ORGANIC ANALYSIS.

QUANTITATIVE

CHEMICAL ANALYSIS.

PART I.

PRINCIPLES OF QUANTITATIVE ANALYSIS.

THE BALANCE. GENERAL PRELIMINARY OPERATIONS.

QUANTITATIVE ANALYSIS is that branch of Practical
Chemistry which treats of the processes by which we determine the relative amounts of the constituents of a body.
QUALITATIVE ANALYSIS informs us simply of the nature of
these constituents, and teaches us how they may be separated. The latter form of analysis always precedes the
former, for, obviously, we must first know what are the
elements present in a substance before we can proceed to
estimate their proportions.

The methods of Quantitative Analysis are subdivided
under the two heads of *Gravimetric Analysis* and *Volumetric
Analysis.* By means of Gravimetric Analysis we seek to
weigh the known constituents of a substance either in their

elementary condition, or in the form of combinations which admit of exact weighing, and of which the composition is already accurately known. Supposing that we wish to determine the composition by weight of a sixpenny piece : qualitative analysis tells us that the coin is made up of silver and copper. We may determine the proportion of the two metals in the solution of the coin, either by separating them out and weighing them in their metallic state, or we may convert the silver into silver chloride, and the copper into cupric oxide, and weigh the two compounds. Since we know the composition of the silver chloride and cupric oxide, we can calculate the amount of silver and copper respectively contained in them, and in this manner determine the relative quantities of the metals present in the coin. In practice, it is usually found more convenient to estimate the constituents in a body by the aid of combinations of known composition, rather than to attempt to isolate the elements. It is evident, therefore, that a correct knowledge of the proportion in which the several elements are present in these fiduciary combinations is of the highest value to the analytical chemist ; and, further, that the exact determination of the combining weights of the elements becomes to him a matter of primary importance.

But it will be obvious on reflection that we can determine the quantitative composition of the sixpenny piece without directly weighing either the metals, or their combinations with chlorine and oxygen. We might determine the amount of the silver, for example, by ascertaining the quantity of hydrochloric acid required to convert it completely into silver chloride. We know that if we add hydrochloric acid to solution of silver in nitric acid (silver nitrate) we obtain insoluble silver chloride ; and that if we add a sufficiency of hydrochloric acid the whole of the silver will be thrown out of solution :

$$AgNO_3 + HCl = AgCl + HNO_3.$$

Now, if we know how much hydrochloric acid (H Cl) is contained in any given volume of the solution which we employ to precipitate the silver, and if we have the means of recognising the exact point at which the formation of silver chloride ceases, we can calculate from the volume of acid required the amount of the silver, since, as the equation tells us, 36·46 parts of hydrochloric acid are equivalent to 107·93 parts of silver. This is the fundamental principle or *volumetric analysis,* a form of quantitative analysis in which we seek to estimate the amount of a substance from the determinate action of reagents in solutions of known strength, the amount of the reacting substance being calculated from the volume of liquid used. Many examples might be adduced to show the wide applicability of this principle of analysis. Let us suppose that we wish to determine the amount of sodium hydrate in an aqueous solution of this substance. If we add a few drops of litmus tincture to the liquid we obtain a blue colouration, which, on the continued addition of hydrochloric acid, eventually becomes permanently red. The acid combines with the alkali to form common salt, which is without action on the colour of litmus ; the final change in colour shows us that the whole of the sodium hydrate is in combination, and that the acid is in very slight excess. If the strength of our hydrochloric acid solution is known to us, that is, if we can say how many grams of H Cl are contained in 1,000 cubic centimetres (for instance) of the liquid, we can calculate, from the number of cubic centimetres we require to add to the soda solution coloured with litmus before it is permanently reddened, how much sodium hydrate is contained in the alkaline liquid originally taken, from the knowledge that H Cl = NaHO ; *i.e.* that 36·46 grams of hydrochloric acid are equivalent to 40·04 grams of sodium hydrate.

THE BALANCE.

The balance affords the only practicable means of measuring the mass of the various forms of matter contained in a substance. Practically speaking, this instrument consists of an inflexible metallic lever or beam suspended near its centre of gravity on a fulcrum or pivot, the masses to be

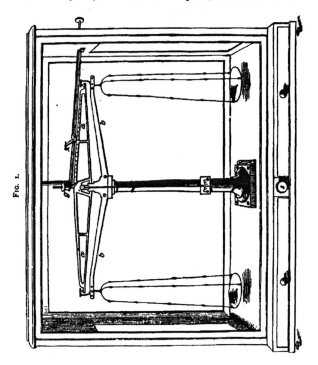

FIG. 1.

compared being also suspended from pivots placed at the extremities of the beam, equidistant from, and in the same horizontal line with, the central fulcrum. For a complete treatment of the mechanical theory of the balance we must refer to special treatises on the subject : in this work we shall mainly confine ourselves to the essential points in its construction and mode of use, and only touch on the mechanical problem in so far as it appears necessary to enable the student to understand the conditions of sensibility and accuracy in the instrument. Fig. 1 gives a representation of a modern chemical balance. The beam *a a* has the shape of an acute rhomboid ; this form of construction combines lightness with inflexibility and strength : on the possession of these qualities in the beam much of the sensibility and accuracy of the balance depends. Through the

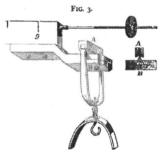

Fig. 2.

centre of the beam passes a triangular piece of hardened steel or agate, termed a *knife-edge*, the lower edge of which turns upon a horizontal plate of polished agate connected with the pillar. At the end of each arm is a similar knife-edge fixed in the reverse position, and bearing an agate plate from which depend steel hooks to hold the wires attached to the pans (fig. 2). These terminal knife-edges are fixed in brass settings, and admit of being adjusted so as to bring them into exactly the same plane with the centre edge. Their relations to the middle knife-edge may be altered by means of the little screws shown in the

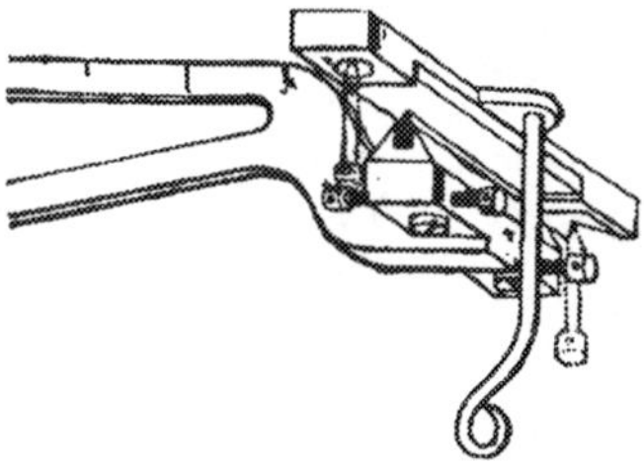

Fig. 3.

figure. Various other methods of arranging the terminal
knife-edges and pan-suspensions have been proposed. Fig. 3
represents a form adopted by continental balance-makers.

As the efficacy of the instrument depends to a large
extent on the preservation of the sharpness of the knife-
edges and the smoothness of the agate planes, it is desirable
to prevent their contact when the balance is not in use.
This is effected by means of the frame *b b* (fig. 1), which
lifts the centre knife-edge about 0·2 millimetre from the
centre plane : at the extremities of the frame are steel
points which enter into little hollows in the lower surface of
the pan-suspensions, and raise them from the terminal
knife-edges. This frame is attached to a rod descending
through the pillar, and connected with an eccentric worked
by a milled-head screw (*s*) situated on the outside of the
balance-case : by means of this movement, the rod, and
with it the frame, can be raised or lowered at pleasure. In
balances of the highest class there is a second eccentric connected with a system of bent levers which carry supports for
the pans ; by means of these supports the pans can be
steadied whilst the weights are being transferred, or their
vibrations can be checked preparatory to releasing the
beam. In some balances the pan-supports are worked by
an independent screw : in others they are worked in conjunction with the movement which raises or lowers the
frame. Where all the movements are controlled by a single
screw these are not made to act quite simultaneously.
When the balance is to be set in vibration, the first action
of the screw lowers the pan-supports ; it next brings down
the centre knife-edge upon the agate plane, and gradually
allows the pan-suspensions to drop simultaneously upon the
terminal knife-edges. For the proper performance of these
movements great nicety of workmanship is needed, for it is
not only requisite that the beam and pan-suspensions should
be properly raised when wanted, but it is also necessary
that the edges and planes should be brought into contact in

a constant position. The movements of the beam are indicated by a vertical pointer which oscillates before an ivory scale fixed to the pillar; this ivory scale is usually graduated into 20 parts, and its middle point or zero is exactly behind the needle when the beam is horizontal. Any inequality in the weight of the arms is compensated by means of a small vane fixed on the top of the beam above the central knife-edge, which may be turned to the right or left as occasion requires. In some balances this compensation is effected by means of little screws travelling along fine threads attached to the ends of the beam (see fig. 3). The stability of the beam is regulated by the aid of a weight termed the *gravity-bob* (*g*) moving along the rod attached to the upper edge of the beam over the centre knife-edge on which the vane works.

In order to protect the instrument from the fumes of the laboratory, and to prevent air-currents from interfering with its action during the operation of weighing, it is enclosed in a glass case, the back, front, and sides of which can be opened at will. The case is supported on levelling screws, by which it can be adjusted to horizontality in accordance with the indications of a spirit-level or plumb-line attached to the instrument. When an object too bulky to be brought within the balance-case has to be weighed, on releasing the screws at the base, the pillar and beam can be turned through an angle of about 60°, so that the ends of the beam project beyond the back and front of the case. The proper adjustment of the beam on the part of the balance-maker is an operation of the greatest nicety. To ascertain if the three knife-edges are in the same plane, he first poises the beam without weights on the pans, and moves the gravity-bob until the vibrations, as indicated by the pointer, become very slow ; he then puts equivalent weights into the pans, and again sets the beam vibrating : if its rate of vibration is unaltered, the adjustment is perfect. If the beam vibrates too quickly or oversets, the gravity-bob is raised or lowered so as to bring the vibrations to the original rate: the number of turns required to effect this is noted, and then the bob is turned in the contrary way through double the number of revolutions, and the slow motion is again produced by means of the adjustments at the ends of the beam. To determine whether the terminal knife-edges are at equal distances from the centre edge, the beam is poised with weights, and the pans, together with their suspensions, are changed from side to side. If the equilibrium is undisturbed, the edges are properly adjusted ; if, however, one side appears heavier than the other, a small piece of bent wire termed a *rider* (see fig. 7) is placed on the lighter side, and pushed along the beam until the equilibrium is again established : the rider is now pushed along half way towards the centre of the beam, and the adjustment made at one end. The knife-edges may be known to be parallel by hanging little hooks upon them and equipoising the beam : on sliding the hooks along the knife-edges, equilibrium should be maintained. The student will better appreciate the skill required in these adjustments when we treat of the circumstances, other than those due to imperfect workmanship, which modify the action of the balance.

We will next briefly state the main conditions upon which the stability, sensibility, and accuracy of the instrument depend.

1. *The centre of gravity of the balance must be situated below the point of suspension,* i.e. the centre knife-edge. If the balance were suspended at its centre of gravity, it would be in the condition of neutral equilibrium, and the beam being once disturbed would have no tendency to reassume horizontality, but would remain in any position given to it ; that is to say, the beam would not oscillate. If, on the other hand, the centre of gravity is above the point of suspension, the beam would be in the state of unstable equilibrium, and would tend to overset with the least preponderating weight.

2. *The terminal points of support* (knife-edges for pan-suspensions) *must be in the same line with the centre point of suspension.* In other words, an imaginary straight line drawn from one terminal edge to the other should just touch the lowest point of the centre knife-edge. If the centre edge is below the line joining the extreme points of suspension, the centre of gravity of the whole system will be raised in proportion as the load is increased, and at a certain point the centre of gravity will be exactly at the point of suspension of the beam, which will then cease to oscillate ; a continued increase of weight will now raise the centre of gravity above the point of support, when the beam will overset. But if the centre edge is situated above the line joining the extreme edges, the centre of gravity of the whole system is lowered in proportion to the increase in load ; that is, its sensibility will diminish with the increase of weight on the pans. When the three edges are in the same plane an increase of weight continually tends to bring the centre of gravity nearer the point of support, but it can never be made to coincide with it : consequently the balance will never cease to vibrate. We thus see why it is necessary that the beam should be perfectly inflexible. If the weights acting on the terminal edges caused them to sink below the level of the centre edge, the centre of gravity of the whole system would become lowered, and the sensibility be lessened. Theoretically, increased weight creates increased sensibility in the instrument : practically, however, this increase in sensibility is more than counterbalanced by other effects of increased weight.

Let *a* be the central knife-edge supporting a beam seen end-on (fig. 4), and *b* and *c* the terminal points of suspension ; the straight line *b c* touches the point *a* of the centre knife-edge. From *a* draw the vertical *a d* ; then the centre of gravity of the beam must be somewhere on this line *a d* : let us suppose it to be at *e.* If equal weights w are suspended from *b* and

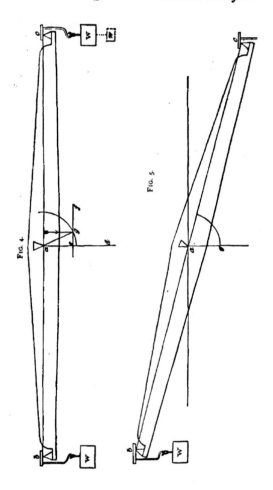

FIG. 4.

FIG. 5.

c, we may conceive that each of the weights acts respectively at b and c : their common centre of gravity falls at a, and the centre of gravity of the whole system of beam and load is somewhere between a and e along the vertical $a\,d$. Since the centre of gravity remains vertically under the point of support, the equilibrium is undisturbed. Suppose now that an additional weight w is made to act at c, the centre of gravity of the combined load is no longer coincident with a, but falls somewhere along the line $b\,c$ in the direction of c, say at f: the centre of gravity of the whole system is at some point along the line $e\,f$, say at g. The centre of gravity is no longer vertically under the point a: accordingly the beam tends to revolve until this condition obtains (fig. 5). The arm $a\,c$ therefore falls, whilst, of course, the arm $a\,b$ is proportionately raised. The angle made by the beam in its new position of equilibrium with the horizontal, due to the preponderance w, is termed the *angle of deviation*: it is equal to the angle $g\,a\,e$. This angle is the measure of the sensibility of the instrument. It is evident, moreover, that (3) *the sensibility of the balance is augmented by bringing the centre of gravity as near the point of support as possible*, whereby g, due to an overplus, becomes proportionately raised towards the line $b\,c$, and the angle of deviation $g\,a\,e$ increased in the direct ratio of the change. If e is so far raised as to be identical with a, then on the addition of the weight w, g would fall on the line $b\,c$; that is, the angle of deviation becomes a right angle, and the beam consequently oversets. The distance between the point of support and centre of gravity in a sensitive instrument probably does not exceed $\frac{1}{400}$ of a millimetre. The arrangement by which the alteration in the centre of gravity is effected has already been described. There is, however, a limit of approximation beyond which, in ordinary work at least, it is practically inconvenient to go, and for another reason than that just adduced. As the centre of gravity nears the point of support, the rapidity of the vibrations of the beam diminishes, and ultimately becomes very slow : the operation of weighing may thus need a greater expenditure of time than is warranted by the degree of accuracy required.

4. *The sensibility of the instrument increases with the length of the arms.* If *a c* be made longer, the distance *a f* will be proportionately increased, and the point *g* will be further removed from the vertical *a d* : the line *a g* will therefore form a greater angle with *a e* ; that is, the angle of deviation will be increased. Here, too, in practice, we quickly find the limit to the length of beam. By increasing the length of the beam we increase its weight, and by increasing its weight we diminish its sensibility.

5. *The sensibility is dependent upon the lightness of the beam.* We may conceive that the weight 2w + *w*, acts at the point *f*, and that the weight *x* of the beam acts at *e*. The position of the centre of gravity *g*, along the line *e f*, is obviously dependent upon the relation of the weights acting at *e* and *f* : if *e x* = (2w + *w*) *f*, then *g* will be equidistant from *e* and *f*, and the smaller *x* becomes in proportion to 2w + *w* the further will *g* be removed from *e*, and therefore the greater will be the angle of deviation. With the view of increasing the sensibility by diminishing the weight, other substances than brass have been proposed as the material of balance-beams, and beams have actually been constructed of aluminium, which possesses a specific gravity of only 2·6, less than one-third that of brass.

6. *The sensibility of the balance is also affected by the friction between the knife-edges and agate planes.* The immediate effect of this friction at the terminal knife-edges is practically to vary the length of the arms. Let us suppose that the impediment to the free motion of the planes of the pan-suspensions is at its maximum, or, in other words, that the planes of the pans are maintained parallel to the direction of the beam during the oscillations of the instrument. Such an instrument would be perfectly useless as a balance, for as one arm was depressed by the action of a preponderating weight, the heavier pan would be thrown inwards,

but its tendency to move would be counterbalanced by the other pan being thrown correspondingly outwards. This variableness, practically speaking, in the length of the arms may be perceived in balances of which the edges have become blunted by wear. Supposing that the width of the knife-edges is x, and the distance between their middle points is y, a glance at the figure (fig. 6) enables us

Fig. 6.

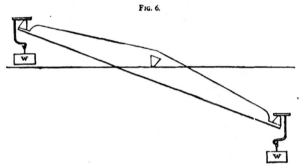

to see that the real lengths of the arms must be $x - y$ and $x + y$. Therefore two loads possessing the ratios of $x + y$ and $x - y$ will be in apparent equilibrium. It is said that some balances will indicate one part in a million; if such an instrument possessed a 20-inch beam, the width of the knife-edges cannot exceed $\frac{10}{1000000}$ of an inch (an inappreciable amount even under a microscope), and the two arms must be adjusted to equality within this length.

When properly adjusted, a balance should satisfy the following tests :—1. When the beam is released the pointer should coincide with the zero on the scale, or make slow excursions tending to equality of amplitude on either side. 2. The equilibrium should not be disturbed when the pans are removed. 3. If possible, the position of the beam should be reversed, so as to cause the arm which points to the right to point to the left, and the beam again be made to oscillate : it should vibrate exactly as before, and finally acquire a horizontal position. Imperfect workmanship in the middle knife-edge or in the planes on which it works is immediately indicated if the beam now behaves differently. In a good balance the centre knife-edge, whether of steel or of agate, is made in one piece, and runs through a perforation in the beam : if, as is occasionally the case, the knife-edge is made in two parts, one being affixed to each side of the beam, it becomes almost impossible to bring the edges into exactly the same straight line. 4. Inequalities in the lengths of the arms may be detected by loading the pans, after the beam has been found to be in equilibrium, with counterpoising weights, and transferring the weights from one pan to the other ; if equilibrium is retained, the lengths of the arms are equal. A final test of the efficacy of the balance consists in weighing an object several times in succession with the greatest exactitude which the instrument admits of : if it is trustworthy the greatest differences will not exceed 0·2 milligram.

It is very desirable that the student, so soon as he has had a little practice in weighing, should make himself thoroughly acquainted with the capabilities of the instrument which he employs. An hour's careful observance of its behaviour under varying conditions will materially conduce to accurate weighing, and save much subsequent time. He should, in the first place, determine for himself the degree of its sensitiveness by accurately noting the deviation from the zero-point, which an overweight of a milligram effects ; (1) when the balance is unloaded, and (2) when carrying 50 grams on each pan ; and (3) when carrying 100 grams. He should at the same time observe the variation in the rates of oscillation as the load increases. As the balance approaches equality, after some experience he will readily be able to estimate the weight required to establish equilibrium, from the extent and rapidity of the oscillations.

The balance-room should have a northerly aspect, and it should not be liable to great or sudden variations in temperature. If possible, the instruments should be so arranged within the room, that in weighing the light falls over the right shoulder of the operator. The position of the several balances should be fixed, and they should be moved as seldom as possible, otherwise their adjustment to horizontality will be continually disturbed, and the shaking of the beams and pans will inevitably interfere with the constancy of their indications. Frequent attempts at readjustment by inexperienced hands will certainly disturb the regular action of the instrument. If, on commencing to weigh, the balance is found not to be in equilibrium, the beam and pans should be lightly brushed with a camel's-hair brush, and the horizontality of the beam again tested. It should be carefully borne in mind that none of the adjustments ought to be disturbed without sufficient reason, and only after a proper inspection of the several parts of the instrument. In a laboratory where the same balance is used by several operators, the necessary adjustments should be invariably made by the assistant, for no one can have confidence in the indications of an instrument which is liable to hasty and improper alteration. A balance suffers more from imperfect preservation than from proper usage. The fumes of the laboratory should be carefully excluded from the balance-room, and the student should never neglect to securely close the doors of the balance-case when he has finished weighing. Careful exclusion of acid fumes will do much to prevent the corrosion of the polished knife-edges and suspensions : if rusting commences at any one point it will rapidly extend over the entire surface of the steel. In order to diminish the humidity of the air a small dish containing dried carbonate of potash or a piece of freshly-burnt lime should be kept within the case. A balance in constant use will require cleaning about every three or four months. The instrument should be carefully taken to pieces, and the loose parts dusted ; the beam, pans, and suspensions should be rubbed with a piece of soft leather, and the movements cleaned and oiled. Occasionally the agate planes will require repolishing by the maker, as the constant working of the knife-edges wears a minute groove in them, easily perceptible by a lens. Lastly, all delicate instruments should be encased in a well-fitting baize or linen bag when not in frequent use ; this will greatly tend to keep out dust and acid fumes.

THE WEIGHTS.

Since the chemist mainly concerns himself with the relative weights of the constituents of a substance, it is, for the greater number of his operations, a matter of indifference what unit he adopts. He needs merely to assure himself that its multiples and submultiples actually have the values which are assigned to them. Experience shows, however, that it is highly desirable that a common unit should be adopted by chemists, and that it should be directly connected with some national standard. Accordingly, we employ almost exclusively the metric system, of which the gram is the standard weight. The reasons which have conduced to the adoption of this system are (1) that the unit is moderately small ; (2) that its multiples and submultiples are derived from it by decimal multiplication and division, *i.e.* by the simplest possible system ; and (3) that it bears a very simple relation to the measures of capacity and length. A set of weights extending from fifty grams to a milligram will be found most generally useful. Such a set should contain pieces of the following denominations :

grams	grams	gram	gram	gram
50	5	0·5	0·05	0·005
20	2	0·2	0·02	0·002
10	1	0·1	0·01	0·001
10	1	0·1	0·01	0·001
	1			0·001

the entire twenty-two pieces making up 101 grams. The

larger weights, down to one gram, are kept in receptacles lined with velvet to prevent their being scratched ; the smaller ones are also kept in separate compartments covered by a plate of glass. The box should be furnished with small forceps, with which the weights are invariably to be handled, as they should never be touched by the fingers (fig. 7).

Fɪɢ. 7.

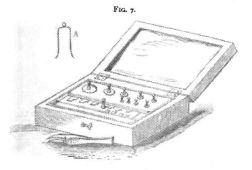

Several substances have been proposed as the material for weights ; for example, rock-crystal and platinum have actually been used by reason of their durability. In general, how-ever, only the smaller weights are constructed of platinum, the others being made of brass gilded by the electrotype process. If the gilding is of moderate thickness, and the weights are preserved with due care, they will be found to be practically unchanged even after many years' use.

Probably the most convenient shape for the brass weights is that of a short frustum of an inverted cone, to the base of which a handle is attached (fig. 8). Beneath the handle is a small cavity containing the minute pieces of foil or wire required to adjust the weights. The smaller weights are preferably made of pieces of stout platinum foil, one corner of each being turned up for holding in the forceps ; the very

Fɪɢ. 8.

C

small ones are occasionally made of palladium or aluminium. All the pieces should have their values plainly stamped upon them, and the compartments of the smaller weights should be large enough to admit of their being easily withdrawn; otherwise the foil will soon become bruised, and the denomination of the weight rendered indistinct.

The greatest care should be taken to preserve the weights from the action of acid fumes. The lid of the box in which they are contained should be tightly fitting, and the box itself should be kept in a bag of soft leather. No attempt should ever be made to clean the weights by rubbing; any dust may be removed by a camel's-hair brush. If by chance the small platinum weights become dimmed or soiled, they may be brightened without injury by holding them for a moment in the flame of the Bunsen lamp. Brass weights ungilded generally become heavier by corrosion; when gilded they usually become lighter by wear. Since this action occurs only at the surface, which increases as the square of the diameter, whilst the mass increases as the cube, it follows that the smaller weights become more quickly erroneous than the larger. The error thus caused by age is, however, exceedingly minute. The slight tarnish is so excessively thin that it requires a very delicate instrument to detect its influence.

We strongly recommend the operator to test his set of weights; for however carefully the remainder of his quantitative work may be done, if he weighs with grossly inaccurate weights, his results, if not valueless, will at least be. inexact. The exact determination of the values of the separate pieces in a set of weights in terms of one of them selected as a standard is an operation of some skill, and requires a considerable expenditure of time. As a rule, the weights of the best makers possess greater accuracy than is required of them in ordinary quantitative processes; nevertheless their examination should never be omitted. The readiest method of detecting errors in the values of the denominations is to place oné of the gram weights on the pan of a delicate balance, adjusted to perfect equilibrium, and equipoise it with pieces of brass or small shot, and finally tinfoil. The weight is then removed and replaced by the second gram weight, and the balance caused to oscillate. If the excursions on either side of the zero are of equal amplitude, the weights are equivalent ; if not, the deviation must be noted. It should not exceed 1 division of the scale from the zero-point. The third gram weight is next tried in the same way, and it is then replaced by platinum weights of the smaller denominations to make up 1 gram, and the balance again caused to oscillate, every deviation from the equilibrium being carefully noted. In the same way the 2 gram piece is compared with two of the single grams, the 5 gram piece with the $2+1+1+1$ gram pieces, and each of the 10 gram pieces with the $5+2+1+1+1$ gram pieces. The larger pieces also should not show greater variations than 1 division from the zero, since the value of 1 scale division with a heavy load on the pans is almost invariably greater than with a diminished load, for the reasons already given. Thus in a certain balance tested by the author, 1 scale division, with no load on the pans, was equivalent to 0·29 milligram ; i.e. a preponderance of 0·29 milligram would cause a variation of 1 division from the zero-point. Under varying loads the value of 1 division on this balance was as follows :—

Load on each pan grams	Value of 1 scale division milligram
10	0·32
25	0·35
50	0·39
100	0·51

The process of testing should be repeated from time to time, particularly when the weights begin to be tarnished. It is also desirable that the student should be able to compare his weights with a standard set.

THE OPERATION OF WEIGHING.

There are three methods of determining the weight of a body by means of the balance: (1) by the process of *direct weighing*, (2) by that of *reversal*, and (3) by that of *substitution*. As the first is the most expeditious, it is the one usually employed, although it is not the most accurate method. The substance to be weighed is placed on one pan, conveniently that on the left, and the weights are placed on the other. The pan originally selected to contain the object to be weighed should be used for the same purpose subsequently. The effect of any inequality in the length of the arms is thus in a great measure obviated. Supposing that we wished to determine the amount of copper in a piece of the pure metal, and that for the purpose of the estimation we wished to weigh out 1 gram of the metal, and that the arms of our balance were of unequal length, the one on the left being 99 millimetres, whilst the other on the right was 101 millimetres long. We place a 1 gram weight on (say) the *left*-hand pan, and counterpoise it with copper. Since two masses acting on a lever are in equilibrium when the products of the weights into their distances from the fulcrum are equal, it follows that the weight of copper equivalent to the 1 gram weight would be 0·9802 gram, for $99 \times 1·0 = 101 \times 0·9802$. The metal is next dissolved and the copper precipitated, these operations being so carefully performed that nothing is lost. The precipitated copper is then weighed, and by mischance it is brought on the left-hand pan; when counterpoised with the weights, it would appear to weigh 0·9705 gram, since $99 \times 0·9802 = 101 \times 0·9705$: accordingly the pure copper would appear to contain only 97·05 per cent. of metal. Had we invariably employed the same scale for the same purpose, we should have determined what we required to know—viz., that the copper contained 100 per cent. of the metal—although we might not know (which would be perfectly immaterial) what was the true weight of the metal employed in the analysis. And by similar reasoning we see that, in the analysis of a substance containing any number of constituents, we ought to use one and the same pan as the object-pan; the results, being strictly comparative, are thus independent of the imperfection of the instrument, since the apparent weights of all the bodies weighed are altered in exactly the same ratio.

We can, however, obtain the absolute weight of the copper taken for analysis, either by the method of *reversal* or by that of *substitution*. In the method of reversal (known as Gauss's) the object (the copper for example) is weighed first in one pan, and then in the other. If the weights are identical, the true weight of the object is at once given. If the weights are unequal, their geometric mean will be the true weight; this is found by multiplying the apparent weights together and taking their square root. Practically the common arithmetic mean of the two weights will be sufficiently accurate, unless their difference is considerable.

In the method of substitution (due to Amiot *), the body to be weighed is first accurately counterpoised; it is then removed, and equilibrium again established by placing weights in its stead. Obviously the absolute weight of the body is expressed by the value of the weights substituted. In practice this method of weighing may be facilitated by using one of the larger weights. heavier than the body to be weighed, as a counterpoise, and adding weights to the object until equilibrium is established. The object is then removed, and weights substituted until the balance is again in equilibrium. The substituted weights express the real weight of the object.

Theoretically these methods are faultless; practically they are subject to at least two minute errors—one due to the impossibility of maintaining the edges of suspension in a perfectly uniform position on the plates whilst the beam is being repeatedly released and arrested, the other due to

* Known also as Borda's.

insensibility. As a balance, however sensitive, requires *some* weight to make it turn, a difference equal to this turning weight may exist between weights apparently equal. The error due to insensibility may be eliminated by weighing the object several times in succession, since the balance ought to turn as readily one way as the other.

The weights should be placed on the *weight-pan* methodically, and not taken haphazard from the box. A very little experience is sufficient to tell, roughly speaking, the weight of an object. Let us suppose that we wished to determine the weight of a platinum crucible which we afterwards found to be 26·715 grams. We place on the weight-pan the 20 gram piece and release the beam—this proves too little; we add one of the 10 gram pieces—this is too much ; we substitute the 5 gram piece for the 10 gram—the weight is again too little; we add the 2 gram piece—this is again too much; we substitute one of the 1 gram pieces for the 2 gram piece —the weight is too little ; we add the 0·5 gram, still too little; also the 0·2 gram, still too little, although we observe that the rate of the vibration of the beam becomes much slower —the balance is rapidly approaching equilibrium. It is at this point that the skill of the operator comes into play ; an expert weigher, familiar with the indications of his instrument, can almost intuitively tell, from the extent and rapidity of the vibrations, what weight is required to establish equilibrium. Until this experience is acquired it will be well to try the addition of the remaining weights in the methodical manner above indicated. We find that 26·7 grams is not quite sufficient to equipoise the crucible ; we add 0·1 gram—too much (the pointer swings with increased energy in the opposite direction) ; substitute 0·05 for the 0·1 gram—still too much ; try 0·02—the pointer vibrates much more slowly, but still indicates that the weight is too great ; we substitute 0·01 gram for the 0·02 gram—the pointer swings with the same slowness, but shows an equal deflection to the opposite side of the zero ; we add ·005 gram—the pointer now makes excursions of equal amplitude—the balance is in equilibrium.

Having assured ourselves that such is the case by reading
off the position of the pointer at the end of the vibrations
several times in succession, we finally arrest the motion of
the beam, and determine the aggregate value of the weights
employèd, (1) from the vacant spaces in the box, and (2)
from the denominations on the weights themselves. This
double reading should never be neglected : the one method
serves to control the other.

It must be carefully borne in mind that whenever an
exchange of weight is necessary, the motion of the beam
must be arrested : *under no circumstances must anything be
removed from the pans when the balance is free to oscillate.*
Neglect of this precaution will quickly ruin the instrument.

The difficulty of transferring and reading the weights is
greatly increased when we arrive at such minute fractions as
the milligram. In order to obviate the inconvenience in-
separable from the use of a number of small weights, which
are apt to be lost or erroneously read, Berzelius proposed
to estimate the last minute fractions (the milligrams and
tenths of a milligram) by a small movable weight sliding
along the beam. A piece of brass gilt, aluminium, or
palladium wire, weighing exactly 1 centigram, is bent in the
form represented in A, fig. 7 ; this weight is called a *rider*,
and by means of the movable rod (fig. 1) worked from the
outside of the balance case it can be placed on one arm
of the beam (that to which the weight-pan is suspended) at
any required distance from the centre edge. The arm from
the centre to the terminal edge is divided into ten equal
parts, and each of these is (generally) subdivided into five
equal parts. The rider weighs exactly 1 centigram or 10
milligrams when placed just above the terminal knife-edge.
When acting at a point on the arm exactly midway between
the two edges, it exerts only half this effect—in other words,
it now becomes equivalent to 5 milligrams ; when acting at
a fourth of the distance from the centre edge, it is equal to
2·5 milligrams, at three-quarters the distance 7·5 milligrams,
and so on. The employment of this very simple contrivance

is attended with much economy of time and trouble, and with greater accuracy, as the rider allows the minutest variation to be estimated, and diminishes the chance of error in reading the weights.

The subdivisions of the centigram may also be rapidly estimated by means of the arrangement represented in fig. 9.

Fig. 9.

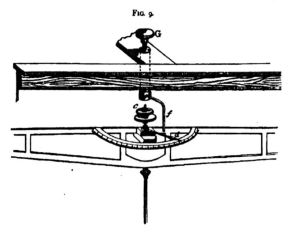

To the bottom end of the rod carrying the screw *c* is fixed a small movable pointer *d*, which moves over a graduated arc. The pointer can be pushed along the arc by means of the arm *f*, worked by the milled-head screw G, placed on the outside of the balance case. This pointer acts as a small weight, its value depending on its proximity to the plane of the beam. The balance-maker effects the graduation of the arc by placing a centigram weight upon one pan, and moving the index in the direction of the opposite pan until equilibrium is established; one milligram of the weight on the pan is now removed, and the position of the index along the arc again noted when equilibrium is re-established. Successive milligrams are removed, and the position of the pointer required to bring about equilibrium is repeatedly determined. The divisions are then subdivided into tenths, each representing o·1 milligram.

The following general rules may prove serviceable to the student in weighing :—

1. Before commencing to weigh, see that the rider hangs upon the projecting pin of the sliding rod, and not upon the beam.

2. Next test the equilibrium of the balance by cautiously lowering the supports and setting the beam in oscillation. If the balance is not in equilibrium, seek for the cause of disturbance, and brush the pans, &c., with a camel's-hair brush, and again test the equilibrium before attempting to move the vane or alter the terminal screws.

3. When it is necessary to arrest the motion of the beam, or to transfer any of the weights, or to remove the body weighed, the supports should be raised the moment that the pointer is opposite the zero of the scale. The screw regulating the eccentric should be gently turned as the pointer travels towards the zero, so that immediately they are in coincidence, a very slight but rapid turn may arrest the beam without any jerking or vibration. If the screw is suddenly turned when the pointer is at the end of an excursion, the beam will be jerked, and its original position of equilibrium, as shown on the ivory scale, will inevitably be disturbed. Carelessness in arresting the oscillation of the beam greatly interferes with the uniform behaviour of the instrument, and with inexperienced operators is the most frequent cause of disarrangement.

4. All shaking of the room or table on which the balance stands should be carefully avoided. The operator should be seated so as to have the ivory scale in direct line of vision, but he should also be able to remove the weights, &c., readily, and to work the sliding rod carrying the rider.

5. If the balance is very nearly in equilibrium, it occasionally happens that it refuses to vibrate immediately after

releasing the beam, and the pointer remains stationary. By gently wafting the air down upon one of the pans, vibration to the required extent may be readily set up. In very accurate weighing it is not always desirable to reopen the doors of the case for this purpose : vibration may then be brought about by *gently* touching the beam or top of the rider with the pin of the sliding rod.

6. As a general rule, the substance to be weighed should never be placed directly upon the object-pan, but should be contained either in a crucible or on a watch-glass. Owing to its hygroscopic nature paper cannot be used if the exact weight of the body is of importance (and when it is not ordinary scales should be used instead of the chemical balance).

7. A body should never be weighed when warm. By placing a warm object on the pan the indications of the instrument are affected to a marked extent. The ascending air-current produced acts against the object-pan and beam above it, and the body appears to weigh less than it ought to weigh. The warm air, after a time, also affects the portion of the beam against which it strikes, and by increasing its length disturbs' the equality in the arms. All substances condense upon their surface a certain amount of air and moisture, the weight of which depends upon their temperature. From this cause also the weight of a body when warm is always less than when cold. A silver crucible, weighing when cold 38·880 grams, was heated over a lamp, and placed whilst hot on the pan ; it now appeared to weigh only 38·835 grams : weighed again when cold, it regained its original weight. If the crucible had contained 0·5 gram of a body to be heated, a loss of 0·045 gram, or 9 per cent., would have apparently occurred, when in fact the weight of the body might have been unchanged.

A platinum crucible when rubbed with a dry cloth and immediately weighed always weighs sensibly less than after half an hour's exposure to the air of the balance case, owing to the condensation of the air upon its surface. It is advisable, therefore, to allow the crucible, if freshly wiped, to remain upon the balance pan or in the case some little time before being weighed.

8. Hygroscopic substances must be weighed in well-covered crucibles, or in stoppered tubes, or between watch-glasses. Liquids must be weighed in covered vessels or in stoppered flasks. Expedients required in particular cases will be mentioned hereafter.

Within certain limits, the deviation from the horizontal in a balance is proportional to the weight causing it. Advantage may be taken of this fact to estimate the last fractions of a weight (i.e. the parts of a milligram) with great accuracy. When the balance is very nearly in equili-brium, it is caused to oscillate, and the position of the pointer when at rest determined from successive observa-tions of the extreme points of the vibrations. So soon as the excursions of the pointer fall within a certain limit, their extent commences to decrease at a regular and uniform rate. Let a_1, a_2, a_3 . . . . . be the extreme points con-secutively reached by the pointer in its oscillations ; then the equilibrium of the balance x is

$$x = \frac{1}{n}\left(\frac{a_1+a_2}{2} + \frac{a_2+a_3}{2} + \ . \ . \ . \ . \ \frac{a_n+a_{n+1}}{2}\right)$$

If we know the weight corresponding to a given deviation from the zero, the estimation of the minute fraction required for exact equilibrium becomes an easy problem. We have first to determine the values of one division of the graduated scale (i.e. the weight required to make the pointer deviate one division) for varying loads. The balance, having been adjusted as nearly as possible, is made to oscillate, and the extreme positions of the pointer in its excursions observed through a telescope. An odd number (conveniently seven) of successive readings are made so soon as the pointer reaches division 6 on the scale : the arithmetical mean of the half-differences between consecutive pairs of observations gives the position of rest of the pointer along the graduated scale.

In an actual determination of the position of rest (x) the following readings were made. The deviations of the pointer from the zero are marked $+$ when they occur to the left of the observer, and $-$ when they occur to his right:

$$
\begin{aligned}
&+ 5\cdot5 \\
&- 4\cdot5 + 0\cdot50 \\
&+ 5\cdot0 + 0\cdot25 \\
&- 4\cdot0 + 0\cdot50 \\
&+ 4\cdot5 + 0\cdot25 \\
&- 3\cdot5 + 0\cdot50 \\
&+ 4\cdot0 + \underline{0\cdot25} \\
&\qquad x = \frac{2\cdot25}{6} = + 0\cdot375
\end{aligned}
$$

An overweight of $0\cdot5$ milligram is then made to act on the beam, the balance again set in oscillation, and successive readings again taken :

Example :
$$
\begin{aligned}
&+ 8\cdot5 \\
&- 2\cdot0 + \quad 3\cdot25 \\
&+ 7\cdot5 + \quad 2\cdot75 \\
&- 1\cdot0 + \quad 3\cdot25 \\
&+ 6\cdot7 + \quad 2\cdot85 \\
&- 0\cdot7 + \quad 3\cdot00 \\
&+ 6\cdot2 + \quad \underline{2\cdot75} \\
&\qquad x = \frac{17\cdot85}{6} = + 2\cdot975
\end{aligned}
$$

The overweight is then removed, and the position of equilibrium again determined : the second determination usually differs to a slight extent from the original one, owing to unavoidable variations in the relative positions of the plates and edges. The mean position is therefore taken as the true point. In the case cited the second determination gave $+ 0\cdot260$: accordingly the mean point is $+ 0\cdot317$. Then the deviation due to the overweight of $0\cdot5$ milligram would be $2\cdot975 - 0\cdot317 = 2\cdot658$ divisions; or 1 division of the scale would be equivalent to $0\cdot188$ milligram (δ). An overweight of 1 milligram is next caused to act on the beam, and the balance is again made to vibrate. This weight ought to produce double the amount of deviation caused by the 0·5 milligram : if any difference is observed, the mean of the two observations is taken as representing the true value of δ. The determination of δ must be made with varying loads, for, as already explained, the sensibility of a balance is seldom constant. In the instrument which gave the foregoing readings the sensibility increased with the load :

Load grams	Value of δ milligrams
0·	0·209
10	0·202
50	0·188

It would obviously be absurd to employ such a refined method of weighing unless we are assured that the differences in the relative values of the weights we use fall within the errors of observation. However good a set of weights may be, the values of the several pieces are never in exact accordance with their denomination : the 50 gram piece, for example, is seldom if ever exactly fifty times the weight of each of the 1 gram pieces, nor has each of the 10 gram pieces exactly ⅕th of the weight of the 50 gram piece. The method of determining minute weights by vibrations, as above described, affords a simple means of comparing the pieces in a set of weights, and of estimating their true values. A delicate balance, placed in a room as little subject as possible to vibrations and changes of temperature, is carefully adjusted, and the value of δ determined on it in the manner already described—(1), with the pans empty ; (2), with a load of 10 grams ; (3), with one of 20 grams ; and (4), with one of 50 grams.*

---

* In these determinations the greatest care is necessary to preserve the balance under perfectly uniform conditions. The operation should be conducted in a room (best in a cellar) set apart for the purpose. If the instrument is exposed to a sudden change of temperature, its equilibrium will almost inevitably be disturbed, owing to the unequal expansion of its arms. A rise of temperature also affects its sensibility (1) by increasing the distance between the centre edge and centre of gravity and (2) by flexure of the beam. The value of one scale division on a delicate balance is invariably greater in summer than in winter.

A slight mark' is made with a sharp-pointed needle on one of the 10-gram pieces, best near its number : similarly one of the 1 gram pieces is marked ', the other is marked ". One of the two platinum 0·1 gram pieces and one of the 0·01 gram pieces have each a second corner turned up ; these weights are respectively styled 0·1' and 0·01' gram. The object of these markings, &c., is to enable the weights to be again recognised. One of the weights (say the unmarked 10 gram piece) is considered as normal; this is placed on the pan to the right of the operator, and is tared with a piece of the same denomination from a similar set. The beam is then cautiously released and made to vibrate, the excursions of the pointer being observed through a telescope placed at a convenient distance (10 or 12 feet) from the balance. The position of rest x is deduced from the readings in the manner already described. The 10 gram is then replaced by the 10' gram piece, the balance is again caused to oscillate, and a second set of readings taken. The adopted standard is again placed on the pan, and a third set of observations made, and again the 10' gram piece is substituted and another set of readings taken. From the mean position of rest (x') deduced from the series with the 10' gram piece, we determine the direction and extent of its difference from the adopted standard, i.e., the unmarked 10 gram piece. The following mean results of actual readings will serve to show the degree of uniformity which may be expected :

Tare v. 10 gram.		Tare v. 10' gram.	
Observation	x	Observation	x'
I	+ 3·09	II	+ 3·10
III	+ 3·09	IV	+ 3·16
V	+ 3·09	VI	+ 3·11
VII	+ 3·09		
Mean	+ 3·09	Mean	+ 3·12

It appears, therefore, that the 10' gram weight is 0·03 of a division heavier than the normal weight. Under a load of 10 grams one scale division on this balance corresponded to

0·32 milligram; accordingly the 10′ gram piece weighs 10+0·00032 × ·03 grams, or 10·00001 grams when the 10 gram piece is taken as normal. The set 5+2+1+1′+1″ gram pieces is in like manner compared with the standard 10 gram piece. The following mean readings were actually obtained:

Tare *v.* standard 10 gram	Tare *v.* 5+2+1+1′+1
x	*x*
I + 3·08	II + 2·85
III + 3·11	IV + 2·82
V + 3·09	VI + 2·90
VII + 3·10	VIII + 2·82
Mean + 3·09	Mean + 2·85

Hence it appears that the series 5+2+1+1′+1″ is lighter than the standard by 0·24 of a division, and accordingly weighs 10·—(0·00032 × 0·24) or 9·99992 grams. The 5 gram piece is then repeatedly compared with the 2+1+1′+1″ series, and the 2 gram piece with the 1+1′ and 1+1″ and 1′+1″ pieces, and so on, the higher and lower denominations being compared in exactly the same manner. The results of the comparisons are thrown together in a table which should accompany the set of weights: in using these the sum of the corrected values of the several denominations is taken as the true relative weight of the object weighed. This table may conveniently resemble the one annexed, which contains the results of a comparison of a remarkably good set of weights by Staudinger of Giessen.

D=Denomination of weight.

W=True relative value.

D	W	D	W	D	W	D	W
100	99·99971	5	5·00002	0·5	0·50002	0·02	0·02002
50	49·99971	2	1·99997	0·2	·19997	0·01	0·01001
20	19·99989	1	0·99995	0·1	·10001	0·01	0·01002
10	10·00000	1′	0·99998	0·1′	·09999	0·01 *	0·00996
10′	10·00001	·1″	1·00000	0·05	·05001	0·01 *	0·01004
					* Riders		

The determination of the weight of a body with the greatest attainable accuracy is a problem of no slight difficulty. It not only demands on the part of the operator considerable skill and a thorough acquaintance with his instrument, but also the knowledge of certain numerical data, some of which indeed can only be approximately known to him. Every substance immersed in a fluid is apparently diminished in weight by the weight of the fluid displaced ; and since all our weighings are made in the fluid which everywhere surrounds us, viz., the air, a balance carrying two weights in equilibrium simply shows that the weight of the body weighed less the weight of the air which it displaces is equal to the aggregate values of the weights less the total volume of air which they displace. Since every body displaces so much air as is contained in the space it occupies, it follows that, in order to determine the true weight of an object weighed in air, we must know also :

1. The volume of the body weighed (v).
2. The total volume of the weights (v').
3. The weight of a given volume of dry air under standard conditions (L).

The weight of the object $-v_L=$ the aggregate values of the weights $-v'_L$,

or

the true weight of the object $=$ the aggregate values of the weights $+v_L-v'_L$.

When the volume of the body weighed is equal to that of the weights employed, $v_L-v'_L=0$: in this case only does the balance directly give the true weight of a body. When the volume of the body weighed is *less* than that of the weights, the expression ($v_L-v'_L$) is negative : the apparent weight of the object is greater than its real weight. On the contrary, when the volume of the body weighed is *greater* than that of the weights, the apparent weight is less than the real weight, since v_L is greater than v'_L.

We can determine v and v' either from the linear dimensions of the bodies, or more easily, and more accurately, from their densities. The value of L requires correcting for variation in temperature, pressure, and atmospheric moisture. We have therefore to observe the thermometer, barometer, and hygrometer at the moment of weighing: v and v' are also not invariable but are dependent on the temperature at the time of weighing; we require therefore (when the greatest possible accuracy is desired) to correct for their expansion.

It is seldom necessary to correct the indications of the balance to this extent, since it is only in the estimation of the combining weights of the elements, and in certain physico-chemical determinations that such extreme accuracy is needed. In such cases Table XI. in the Appendix will be found useful. The scope of this work will not permit of a fuller discussion of the precautions and corrections necessary in the exact determination of weight. We would refer for more complete information to a memoir by Bessel, in the 'Astronomische Nachrichten,' vol. vii., or to Schumacher's paper, 'Ueber die Berechnung der bei Wägungen vorkommenden Reductionen' (Hamburgh, 1838), in which the *principles* of the corrections are very fully explained: the physical data, however, need revision. In Kuppfer's work, 'Travaux de la Commission pour fixer les Poids et Mesures de Russie' (St. Petersburg, 1841), is given a full account of the method of vibrations; and, lastly, in Prof. W. H. Miller's classical memoir, 'On the construction of the New Imperial Standard Pound' ('Phil. Trans.' Part III. for 1856), the best manner of conducting Gauss's method of reversal is described; the tables of correction therein given are based on the most accurate data.

GENERAL PRELIMINARY OPERATIONS.

Before we commence any quantitative investigation it is desirable that we should have a clear conception of its object—that we should understand the question our inquiry is intended to settle. If we steadily bear in mind the reason of our labour we shall be guided in the proper selection of the specimen of the substance which we desire to analyse. Supposing that we have a mineral, and wish to determine its composition with the object of elucidating its constitution, we ask ourselves in the outset—would an analysis of this particular specimen afford a proper solution of the question ? We examine it with a lens, or by some other appropriate means, to learn if it is free from foreign matter ; if it is imbedded in a matrix, we carefully remove the adhering gangue ; we then break up the mineral, and select the cleanest and apparently purest pieces ; in short, we do everything that the nature of the case suggests to assure ourselves that we have an individual body to analyse, and not a mixture of substances.

Again let us suppose that we are called upon to examine a cargo of ironstone, or other heterogeneous mass, with the view of ascertaining its value. We should not content ourselves with examining the first lump of the mineral which came under our notice, but we should carefully select and mix a sufficient quantity from various parts of the mass, reduce the mixture to coarse powder, thoroughly intermix it, and then take a portion for analysis.

In the plan of instruction given in this book, the first work of the beginner in quantitative analysis is the determination of the constituents of simple and definite compounds of which the composition is already established. One of the objects of these exercises is to afford him the means of gauging his progress in manipulative skill. To this end the substances to be analysed must actually contain only those constituents they are represented to contain ; in other words, they must be pure. If their purity cannot be guaranteed, the main object of these exercises is missed : the student is not in a position to compare his experimental results with the supposed composition of the body, and from the want of a sure control he fails to acquire that degree of proper confidence in his work which every operator ought to possess.

Many of the operations of quantitative analysis are of continual recurrence. They may therefore be most conveniently described in this introductory part.

Mechanical Division.—In order to render the substance we wish to analyse more susceptible to the action of solvents or fluxes, it is generally desirable to reduce it to a more or less finely-divided condition. This operation is usually conducted in mortars. The kind of mortar to be employed depends upon the hardness of the substance ; in all cases the material of the mortar and pestle must be considerably harder than the body to be powdered, otherwise the substance to be analysed will be inevitably contaminated with the material of the mortar. In general, smooth porcelain mortars suffice for pounding salts, whilst most minerals require to be powdered in mortars made of agate. The agate mortar and pestle should be free from cracks or crevices : the pestle may be conveniently inserted into a wooden handle, which renders it much easier to use. Very hard substances should first be broken into small pieces by wrapping them in paper, and striking them with a hammer upon a smooth surface of iron. The pieces should then be reduced to coarse powder in the steel mortar (fig. 10).

Fig. 10.

a is a solid block of steel, into the slight cavity at the top of which fits the hollow cylinder *b* ; in this cylinder are placed the pieces of the substance to be crushed. The solid pestle *c* is then placed in the cylinder, and repeatedly struck with a hammer until the pieces are sufficiently broken up. Some minerals which suffer no change on ignition (*e.g.* quartz containing gold) may be disintegrated by being repeatedly heated and thrown into cold water.

In order to obtain the complete decomposition of many minerals and insoluble bodies by the action of fluxes, it is

necessary to reduce them to the finest possible state of sub-
division. This was formerly frequently effected by the
process of *elutriation.* The substance was triturated with a
little water in an agate mortar, and the pasty mass thrown
into a quantity of distilled water. After settling for a minute
or two the turbid liquid was poured into another vessel, and
the subsident portion was rinsed back into the mortar and
again triturated ; this process being repeated until a sufficient
amount of the suspended substance was obtained. After
standing for a few hours, to allow the finely-divided matter
to subside, the supernatant liquid was decanted off, and the
powder thoroughly dried. This process is now less frequently
employed in quantitative analysis than formerly, for the
reason that it is found that very few substances are entirely
unacted upon by water ; even finely-divided felspar and
granite give up a portion of their constituents, and
powdered glass loses weight considerably when thus
treated with water. In the case of mixed substances it very
generally happens that some portions are more easily
reduced to powder than others, and that some have a very
different specific gravity from others ; hence it is always
doubtful if a complex substance after elutriation has exactly
its original composition.

The majority of bodies may be obtained sufficiently
finely divided by patient pounding and careful sifting. The
sifting is best effected through fine cambric or muslin. A
piece of the clean and dry fabric is tied over a beaker, about
10 centimetres in diameter, and the powder is thrown upon
the cover, which is then gently tapped with a glass rod. That
which fails to pass through the cover is again triturated and
sifted, the process being repeated until the entire mass has
passed through into the beaker. The powder is again to be
returned to the mortar in small portions at a time (using not
more than will cover a sixpenny piece), and ground until
every trace of grittiness has disappeared, and the substance
cakes in an impalpable dust round the pestle.

Desiccation.—It has already been stated that before we can proceed to analyse a substance we must be assured that it is free from all unessential constituents. The most frequent of these unessential constituents is moisture, by which term we also understand the water over and above that which may be proper to the constitution of the compound. This mechanically-held water may be due to the method by which the body has been prepared, as in the crystallisation of salts, or it may be derived from the atmosphere, as in the case of certain minerals. The majority of substances require to be dried before they can be analysed quantitatively. The method by which this can be most properly and readily accomplished depends upon the nature of the body. If the substance contains water of crystallisation, repeated pressure between folds of filter-paper often suffices to remove the moisture. Occasionally it will be better to place the finely pulverised body in an artificially dried atmosphere, over some hygroscopic substance, such as calcium chloride or strong sulphuric acid. Fig. 11 represents a convenient form of drying apparatus or *desiccator*, as it is often termed. It consists of a glass bell-jar with ground and greased rim, resting on a plate of ground glass; the dish is partly filled with strong sulphuric acid; the tripod may be made of glass rod, and the circular plate to support the dish or crucible containing the substance of thin wood or metal.

Fig. 11.

It is sometimes desirable to hasten the desiccation by conducting it under diminished pressure; the apparatus has therefore an arrangement to connect it with the air-pump or other instrument for procuring a vacuum.

Fig. 12 represents a more portable form of desiccator; it is especially convenient for allowing hot crucibles, &c.,

FIG. 12.

to cool in a dry atmosphere preparatory to weighing them. The lid and lower portion are of glass, ground together, their perfect conjunction being secured by a slight film of grease. A brass rim, fitting into the aperture of the lower vessel, which contains sulphuric acid or calcium chloride, carries a triangle to support the crucible, &c.

Substances experiencing no alteration in the neighbourhood of 100° may be more quickly dried in the steam-bath.

FIG 13.

Fig. 13 represents a simple and convenient form of this apparatus; it consists of a chamber surrounded on five sides by an outer case of sheet copper, in which the water is placed, and which only communicates with the air at *a* and *b*. Water continually drops into *a* to replenish that lost by evaporation, and the steam makes its escape through *b*. The atmosphere within the chamber may be renewed through the holes *c c* in the door.

If the body bears a higher temperature without change, it may be heated in the air-bath. Fig. 14 represents a very

FIG. 14.

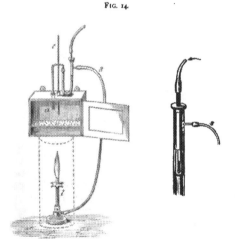

simple form of this apparatus; it is made of sheet-copper, and is heated by the lamp *l*, the flame of which should be surrounded by an earthenware cylinder, indicated by the dotted lines in the figure. The substance to be dried is placed on the shelf within the chamber, the temperature of which is given by the thermometer *t*. In certain analytical operations it is desirable to maintain the bath at a constant temperature for a considerable time ; the flame must therefore be kept of a constant size, and be corrected for variations in the pressure of the gas. It is very convenient when the bath itself can be made to regulate its temperature, so that if it becomes over- or under-heated it can momentarily cut off or increase the supply of gas. The apparatus shown in the figure is fitted with one of the many forms of regulators which have been described. The U-shaped tube contains mercury, into which dips a tube connected with the gas supply ; the gas from this tube passes through a narrow slit, thence up a glass tube, through the side-tube *s*, with which the caoutchouc tube of the lamp is connected. By means of a loosely-fitting screw, the gas-supply tube can be raised or depressed within the mercury, so that the length of the slit, and therefore the amount of gas passing through the apparatus for a given pressure, can be varied by surrounding it with more or less mercury. The other end of the U-tube is connected with a reservoir of air *a*, placed in the bath. If the screw is so regulated as to maintain a given temperature when the reservoir is heated, any increase or diminution of this temperature will be accompanied by a proportional increase or diminution in the volume of air within this reservoir, and a corresponding rise or fall in the height of the mercury surrounding the orifice through which the gas issues to the lamp. By means of this arrangement a uniform temperature within the bath (within 2° at 150°–170°) may be readily maintained.

Weighing out the substance.—Having obtained it in a fit state for analysis, and having fixed upon the scheme of separation to be adopted, the student next weighs off a certain amount of the substance, and proceeds to treat it in accordance with the requirements of his plan. No exact general rules can be given as to the amount which will be required for the analysis, since so much depends upon the nature of the body, and the proportion of its several constituents. The greater the amount taken, the more accurate, *cæteris paribus*, should be the analysis, since the unavoidable errors in precipitating, washing, and weighing do not exercise the same degree of influence on the final result, when the quantity of the substance is large, as when it is small. On the other hand, the smaller the quantity operated upon the sooner will the analysis be finished, but at the same time the demand upon the manipulative skill of the operator will be increased. The object for which the analysis is required can alone tell us how far we should sacrifice accuracy to time. As the student will glean from the following examples, no strictly uniform plan can be given of the manner in which substances should be weighed off for analysis; in general, however, the body, especially when in the state of powder, is weighed out from tubes. The light tube containing the body, and fitted with a good cork, is accurately weighed, the tube is removed from the pan, the cork is withdrawn, and the proper quantity of the substance cautiously shaken out into a beaker or crucible, as the case demands ; the cork is now replaced, and the tube and its contents are again weighed. The difference between the two weighings gives the amount of the body taken for analysis.

The further treatment of the substance depends upon the nature of the constituent or constituents to be estimated. The experience to be gained from the examples which follow will suggest the proper methods. It most frequently happens that the body is to be brought into a state of complete or partial solution, and the constituents separated either by evaporation or by precipitation, or by both processes. Thus we can determine the nitre in gunpowder by treating that substance with water, whereby the salt is dissolved, separating the solution from the undissolved portion, evaporating it to dryness, and weighing the residue. We can analyse common salt when in solution by precipitating the chlorine by the addition of silver nitrate, and weighing the silver chloride produced : the solution still contains the sodium (now as sodium nitrate) ; this, after the removal of the excess of the silver by appropriate means, can be obtained by evaporation. Precipitation can only be resorted to when the precipitate is practically insoluble in the liquid in which it is formed, and when, possessing a constant composition, it admits of being freed from foreign substances, and of being accurately weighed.

Evaporation.—Liquids are generally concentrated by evaporation in porcelain basins placed over a lamp, care being taken that the solution never enters into actual ebullition, as this would occasion loss by portions being projected from the dish. Unless the evaporation is conducted in a room set apart for the purpose, it will be advisable, indeed actually necessary, if many persons work together in the laboratory, to protect the liquid from dust. A piece of glass rod bent before the lamp in the form of a triangle, and covered with a sheet of filter-paper, and supported on a stand over the dish, forms an efficient shield (fig. 17, p. 45). In the evaporation of acid liquids it must not be forgotten that the fumes may dissolve out the inorganic constituents of the paper (iron, lime, &c.) and the condensed vapour dropping back into the dish may contaminate the liquid with those substances. In such cases the paper used must be freed from these soluble matters by treatment with acid in the manner to be hereafter described.

Liquids containing gas, which is evolved in bubbles on the application of heat, are very liable to sustain loss by spirting. In such cases the dish should be covered with a large watch-glass, and the liquid should only be gently heated so long as the evolution of gas continues. When it has ceased, the projected portions may be rinsed from the watch-glass back into the dish. The evaporation of such liquids may also be conducted with less chance of loss in obliquely-placed flasks, which should only be half filled: the portions spirted strike against the upper part of the flask, and are washed back again into the main body of the liquid by the condensing steam. With the flask so tilted the liquid may even be gently boiled, with a very remote chance of anything being projected.

Occasionally a liquid has to be evaporated to dryness in a platinum crucible, in order that the residue may be weighed. If the boiling point is much higher than that of water, the evaporation is best conducted by heating the crucible, placed obliquely, in the manner seen in fig. 15, the heat being directed upon the crucible above the level

Fig. 15.

Fig. 16.

of the liquid. By placing the lid in the position indicated in the figure the evaporation is materially accelerated since a current of air is thus caused to play over the surface of the liquid. A little piece of wire gauze placed on the top of the lamp, enables the smallest flame to be produced without any fear of the gas igniting within the tube. This method of evaporation from the surface is especially serviceable if the liquid contains a precipitate; by heating the crucible at the bottom it is almost impossible in such a case to prevent loss by succussion or *bumping*. It is also useful when the heated solution has a tendency to creep up the sides of the crucible ; the liquid evaporates as it ascends, and meets the heated surface, and the residue is prevented from passing out over the rim. Or the crucible may be placed in a vertical position with the lid partially over the side, so that a small flame placed beneath the lid heats the extreme end to dull redness. By conduction the whole lid becomes hot, and radiates sufficient heat to effect a tolerably rapid evaporation of the liquid.

But as a rule it is safer to conduct the evaporation of liquids over the water-bath. Fig. 16 represents one of the simplest forms of this apparatus ; it consists of a vessel of sheet-copper, about 15 centimetres in outside diameter, partially filled with water, and set over a lamp ; the vessel to be heated by the steam is placed on the top. The bath is furnished with a number of flat rings of various diameters, adapted to receive vessels of different sizes. In order to guard against the effect of inadvertent evaporation of the water in the bath, the apparatus, as represented in the figure, has a simple contrivance for turning off the gas when the copper basin becomes dry. The lamp is provided with a cock, the lever of which is prolonged and weighted with lead : it is kept in position by a piece of thin thread passing over the rim of the basin, and attached to a hook at the bottom. When the basin becomes dry, the thread chars, and breaks, and the lever falls and shuts off the gas.

It is far better, however, so to arrange the bath that the water is continually replenished. The apparatus seen in fig. 17 is designed with this object. The water flows in from the main at *a* ; by raising or depressing the glass tube which slides watertight through a piece of caoutchouc tube slipped over the lower portion of *a*, any required height of water may be obtained in *a*, and accordingly in the bath. The overflow runs through *b*, and may be carried away by a piece of attached caoutchouc tube. This bath has also a number of flat rings, to suit vessels of various sizes.

Bunsen has devised an excellent form of water-bath, which,

when once regulated, necessitates no attention on the part
of the operator in regard to the water supply. It is represented in fig. 18, p. 46.

The bath A is made of sheet copper, and is partially
filled with water, which is heated by the lamp *a*, the flame

FIG. 17.

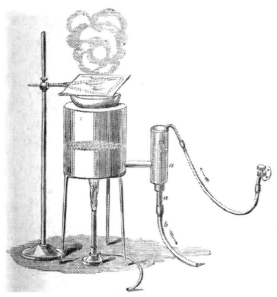

of which passes into the chimney shown in the figure by
dotted lines. The fresh water enters from the apparatus B.
This consists of a wide glass cylinder, nearly filled with
water, in which is a float, the lower cylindrical part of
which contains mercury, whereby it is maintained in a vertical position, and at a certain height in the water. Through
the upper opening dips a tube *f*, connected with the water
supply ; this tube is fastened to the cylinder, but the depth

to which it dips into the float can be so regulated that at
the proper level of water the float rises, and the mercury
cuts off the entrance of the water.   The water-bath is connected with the cylinder by means of the tube e; as the
water evaporates, its level in B sinks, whereby the float falls,
until the end of the tube f is uncovered by the mercury ;
water now enters and flows over into B, and the float, again

FIG. 18.

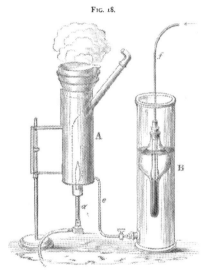

rising, shuts off the water.   This apparatus is more especially adapted to a large laboratory, since any number of
the water-baths may be connected together, one cylinder
and float serving to replenish them all, without waste of water.

It must not be forgotten that the material of the vessel in
which the evaporation is conducted, is, in general, more or
less attacked by the liquid ; it has already been stated that

even pure water dissolves very appreciable quantities of glass. The influence of the matter dissolved from the flasks, &c., used in the operations, is too frequently lost sight of in quantitative analysis : there is no doubt that the results are affected to a greater degree than is usually supposed. Experiment has shown, that, in the case of new vessels, the amount of glass dissolved by a heated liquid is always greatest in the first hour, and gradually diminishes, until it reaches a certain amount, after which the quantity passing into solution is, within certain limits, proportional to the time of action. The amount dissolved is proportional to the surface on which the liquid acts, and is independent of the amount of liquid vaporised, so long as it is maintained at the boiling temperature ; that is, the mere rapidity of the evaporation is without influence. The amount dissolved is in proportion to the temperature of the liquid. 400 c.c. of boiling water in a glass flask of 600 or 700 c.c. capacity* dissolved in the first hour 8·9 milligrams ; in three hours, 14·8 milligrams ; in six hours 22·5 milligrams ; in twelve hours, 32·5 milligrams ; in twenty-four hours, 53·3 milligrams, and in thirty hours, 66·5 milligrams. The same quantity of dilute hydrochloric acid (11 per cent.), boiling in a similar flask, dissolved only 4·2 milligrams in the first hour ; 5·1 milligrams in the third ; 7·3 milligrams in the sixth ; 9·4 milligrams in the twelfth ; and 17·0 milligrams in thirty hours. Dilute hydrochloric acid exerts much less action therefore than pure water. Nitric acid in like manner exerts comparatively little action on glass. 400 c.c. of dilute ammonia (9 per cent.), dissolved from a precisely similar flask 6·7 milligrams in one hour ; in three hours, 15·5 milligrams ; in six hours, 25·3 milligrams ; in twelve hours, 43·9 milligrams ; in twenty-four hours, 84·8 milligrams, and in thirty hours, 99·6 milligrams. It appears that the extent of action of ammonia-water varies very slightly with its strength. A solution of ammonium chloride (7 per cent.), dissolved in one

* Composition of glass, SiO_2 73·8, CaO 8·6, Na_2O 14·0, K_2O 0·60.

hour 4·2 milligrams ; in six hours, 7·3 milligrams ; in fifteen hours, 9·6 milligrams; and in thirty hours, 14·6 milligrams. As a general rule, liquids possessing an acid reaction, even when they contain salts in solution, dissolve less of the glass than when they have an alkaline reaction. The comparatively small quantity dissolved by the ammonium chloride solution is in a great measure due to the fact that this liquid acquires an acid reaction on boiling, owing to the dissociation of the sal-ammoniac : the liberated hydrochloric acid appears to exert a preservative action on the glass. Dilute sulphuric acid, however, exerts a marked action, twice as strong indeed as that of water. The amount dissolved by alkaline fluids is very considerable, even when the quantity of alkali is small ; in six hours 400 c.c. of a boiling liquid, containing 1 per cent. sodium carbonate, dissolved 34·8 milligrams; the addition of $\frac{1}{10}$ of a per cent. of caustic potash to water increases its action threefold. Certain salts, as ammonium carbonate, calcium chloride, common salt, potassium chloride, nitre, act upon glass to the same extent as water ; sulphate and phosphate of sodium act much more energetically, the action of the latter salt being six times that of water. Direct experiment has shown that the glass in all these cases is virtually dissolved ; the liquids do not extract any one constituent in preference to others. Very little difference is observed in the action of the liquids on glass of varying composition, but the true Bohemian glass, rich in silica, and poor in soda, is of all the least attacked. Porcelain vessels are scarcely acted upon by any heated liquids, with the exception of the alkalies, and even in their case the action is very much smaller than with glass : therefore vessels of the former material should invariably be used in evaporations, &c., whenever circumstances permit.*

The precipitation of substances intended to be collected and weighed is usually effected in beakers, on account of

* Emmerling, Ann. der Chemie und Pharm., 150.·257.

the facility with which the bodies may be transferred, either to the filter or to the vessels in which they are to be weighed. In cases where the liquid is strongly alkaline, or where it has to be heated for some time, it is better to use porcelain basins. The separation of the precipitate from the liquid in which it is formed, is accomplished either by *decantation* or by *filtration*, or by a combination of these processes. In general, where the liquid has to be filtered, it is advisable to allow it to stand at rest for some hours after the addition of the precipitant, for the reasons : (1) that the complete separation of the substance in the insoluble form occurs only after some time ; and (2) that in certain cases, if thrown on the filter immediately after precipitation, it is apt to pass through the pores of the paper. Before proceeding to separate the liquid, the operator must invariably assure himself that the precipitant is in excess, by adding a few

_Fig. 19.

drops of its solution, and noting if any further turbidity is produced. The clarification of the liquid on standing allows this to be ascertained with greater certainty. In cases where the precipitate forms only after some time, the precipitant must be added to a portion of the supernatant liquid, poured into another vessel.

The separation of precipitates by decantation is but seldom resorted to, on account of the length of time which it occupies, and the comparatively large amount of water needed for thorough washing. If, however, the precipitate has a high specific gravity, and is practically insoluble in water, as in the case of silver chloride, metallic mercury, &c., the process may be advantageously employed. The subsidence of the precipitate takes place most readily in a vessel of the form seen in fig. 19, p. 49 : this should be made of glass sufficiently thin to be heated without risk of fracture, since warming greatly accelerates the subsidence. The clear liquid is conveniently removed by a syphon, the longer limb of which can be closed by a pinch-cock, so that when the flask is replenished with the washing fluid, the syphon, being always filled with liquid, can again be set in action without the operator being under the necessity of refilling it. The precipitate is then transferred by the aid of the wash-bottle to the crucible or dish in which it is to be weighed, the fluid used in the transference being poured away, as far as practicable, and the precipitate dried. The decanted liquid should invariably be set aside, in order to allow any of the insoluble substance which may have inadvertently been carried over, to subside : if any is detected, it is separated from the liquid in the manner described, and either added to the main quantity or weighed by itself.

In the majority of cases, decantation and filtration are combined ; that is, the liquid is poured through the filter without disturbing the precipitate, and the precipitate, after having been agitated with fresh washing water, is allowed to subside during the time occupied by the contents of the filter in passing through the paper. Filter paper should permit of rapid filtration, and yet possess pores sufficiently minute to prevent the passage even of the most finely divided precipitates ; it should, moreover, be as free as possible from inorganic matter. The Swedish filter-paper, with the water-mark 'J. H. Munktell,' is generally considered to fulfil these conditions in the highest degree. It contains about 0·4 to 0·6 per cent. of ash, consisting of

Silica	Alumina	Iron	Lime	Magnesia
35·16	3·84	45·06	14·09	1·01 = 99·16.*

* From an analysis communicated by Mr. Walter Dearden, Owens College.

The amount and nature of the ash vary, however, with different 'makes' of the paper.

The paper should be cut into filters of various sizes by the aid of circular patterns made of tin-plate or sheet-zinc; these may with advantage have the radii 3, 4, 5, 6, and 8 centimetres. The filters possessing these radii are respectively designated as Nos. 3, 4, 5, 6, and 8. The filters should be treated with dilute hydrochloric acid (which extracts nearly the whole of the inorganic matter, with the exception of the silica), and afterwards be thoroughly washed with water and dried. This treatment with acid may be conveniently made in the apparatus represented in fig. 20. The ready-cut filters are placed in the vessel, and covered with dilute hydrochloric acid (1 part acid to 20 of water), which is allowed to act for a few hours. On opening the pinch-cock at the bottom, the acid liquid flows away; the filters are then to be repeatedly washed with water until every trace of acid has disappeared, after which they may be dried in the water-bath.

FIG. 20.

The operator must now determine the amount of ash left on burning the prepared filters, since this requires to be subtracted from the final weighing of the separated substance. A light porcelain crucible is heated over the lamp, placed in the desiccator and weighed when perfectly cold. The cru-

FIG. 21.   FIG. 22.

cible is placed on a smooth glazed sheet of paper, fig. 21; one of the filters (No. 5, for example) is repeatedly folded over in plaits of about 1 centimetre in breadth, and tightly rolled between the finger and thumb from one end of the folded length to the other until it has the form seen in fig. 22. About half the length of a piece of platinum wire 40 centimetres long is wrapped round the rolled-up filter, which is now lighted at the lamp and held over the crucible. The flame quickly disappears, and the paper becomes reduced to a mass of glowing carbon. As soon as the last spark has died out—*but not till then*—the flame is made to play on the ash held over the crucible, to complete the combustion of the carbon. The ash should now be white or have at most a reddish-gray tinge, without the least trace of blackness. Care should be taken not to heat the ash more strongly than is necessary to burn the carbon, or it may fuse to the platinum wire. This more readily happens with filters which have not been treated with acid, and which, therefore contain comparatively large quantities of lime, iron, &c. The ash is now shaken out of its platinum cage into the crucible, and by tapping the wire against the rim of the crucible, any adhering traces are readily detached. This process is to be repeated with five additional filters; the crucible is again placed in the desiccator, and when cold re-weighed. The difference between the two weighings divided by 6, gives the amount of ash left by a No. 5 filter. Call this amount a: it is easy from this to calculate the ash left by each size of filter. A No. 5 filter has a radius of 5 centimetres; its superficial area is $=r^2\pi$ or $5 \times 5 \times 3.14 = 78.5$ sq. centimetres. Required, for example, the weight of ash of No. 8 filter (x): $8 \times 8 \times 3.14$ area $= 200.9$ centimetres; and $78.5 : 200.9 :: a : x$. It is convenient to prepare a large number of such filters at a time, and to calculate and arrange in a little table for use the amount of ash left by the different sizes.

Glass funnels should always be employed in quantitative analysis: the sides should be inclined at an angle of 60°, and should be free from irregularities or bulgings; the stem should not be too short or too wide, and the end should be cut obliquely. The size of the funnel to be employed of course depends upon the size of the filter required; the filter must never project beyond the funnel; it should be within one or two centimetres from the edge. The size of the filter in its turn depends upon the bulk of the precipitate to be filtered : the precipitate should not occupy more than half the capacity of the filter, or the process of washing will be very tedious.

The filter paper should be carefully folded, and properly placed in the funnel, moistened with hot water (unless circumstances forbid this), and pressed with the finger so as to cause it to fit closely to the funnel; for the better it fits, the more rapidly will it filter, and the less will be the danger of rupture on washing. The funnel is placed in a convenient stand, so that the edge of the stem touches the side of the vessel intended to receive the filtrate. By allowing the liquid to flow down the side, all splashing and consequent risk of loss of the filtrate is avoided. The liquid to be filtered should never be poured directly into the funnel, but down a thin glass rod, the stream being so directed as to fall against the side of the filter; if poured into the apex, loss by splashing will inevitably ensue. The rim of the vessel containing the liquid to be filtered should be *slightly* greased with lard (free from salt); this prevents the chance of the liquid running down the outside of the vessel. Whilst not in use, the rod is placed in the vessel containing the precipitate, or, if this ought not to be disturbed, in a little flask or beaker, which is afterwards rinsed out into the filter so soon as the whole of the precipitate has been transferred. It is advisable to cover the various vessels with glass plates during the progress of the filtration, to prevent dust falling into them. The plate covering the beaker in which the filtrate is received must of course have a small hole at the side to admit the stem of the funnel; this may be readily snipped out by a pair of pliers, or by a key, the wards of which allow of the insertion of the glass.

It frequently happens that small particles of the precipitate firmly attach themselves to the sides of the vessel and cannot be rinsed out on to the filter. To remove them, the end of the glass rod should be covered with a short piece of thin unvulcanised caoutchouc tubing (about 1 centimetre long): by rubbing this against the sides of the vessel, the last traces of the precipitate may, generally, be readily detached. Or, instead of the rod coated with india-rubber, a feather may be used, the plumules of which have been torn away to within 2 centimetres of the end ; those remaining are to be cut parallel to the quill and within ·5 centimetre of it. In transferring precipitates from a basin, the little finger may be often advantageously used to rub away any of the substance from the sides. In all cases it must not be forgotten that the rubbing instrument, after use, must in its turn be carefully rinsed. If the substance cannot be detached by mechanical means, it must be re-dissolved and again precipitated.

The form of wash-bottle best adapted for use in quantitative analysis is seen in fig. 23 ; as the nozzle is moveable,

FIG. 23.   FIG. 24.

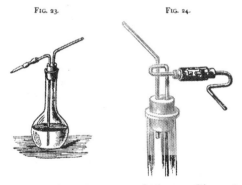

the jet may be directed to any required spot. Fig. 24 shows another kind of wash-bottle with moveable nozzle : it is especially convenient for washing down the precipitate from an inverted beaker held over the funnel. The orifice of the nozzle should not be too large, or the amount of water required to bring a precipitate on to the filter becomes unnecessarily great. In washing a precipitate on the filter, the stream should be directed round the edge of the paper, and care should be taken that the force of the jet is not so great as to rupture the paper. Carelessness in directing the jet will inevitably cause portions of the precipitate to be projected out of the funnel. The operator should also guard against the formation of channels in the mass of the precipitate, through which the water tends to flow without coming into contact with the bulk of the substance. He should never refill the filter with liquid until the previous quantity has passed through : neglect of this rule not only retards the process of washing ; but often occasions a turbid filtrate. As a rule hot water should be employed in washing ; its use accelerates the process greatly ; the few cases in which it is objectionable will be mentioned hereafter. In order that the heated wash-bottle may be conveniently handled, a coil of thick string, or some other badly conducting material, may be wrapped round its neck.

Occasionally the washing is conducted with other liquids than water. Thus in the estimation of potash and ammonia the double platinum salts are washed with alcohol, and in the determination of magnesia and phosphoric acid the precipitate is washed with dilute ammonia water. One separate wash-bottle should be employed for all the special liquids ; as it will be comparatively seldom used, it may have only half the ordinary capacity. By attaching a small piece of caoutchouc tubing to the end of the shorter tube of this wash-bottle beneath the cork, cutting a slit through the caoutchouc to within a centimetre from the end, and stopping it nearly up to the slit with a small piece of glass rod, a simple valve is formed which effectually prevents the escape of the vapour of these special washing fluids—some of which, like sulphuretted hydrogen and ammonia, are very irritating. The valve only opens by inward pressure and closes immediately when this pressure is withdrawn.

We cannot too strongly impress upon the beginner the

necessity of conscientiously performing the operation of washing ; imperfect or careless washing is a very frequent source of error in quantitative analysis. He should invariably ascertain, before he discontinues the operation, that the liquid passing through no longer contains any of those substances which it is the object of the washing to remove : thus in the determination of chlorine as silver chloride, he should test the filtered wash-water by adding to a portion of it, collected apart in a test-tube, a drop of dilute hydrochloric acid ; if the silver chloride has been washed free from the excess of silver nitrate, not the faintest turbidity will be produced.

In certain cases, however, such methods of testing the perfection of the washing are inapplicable. It is obvious that if we wash twice with a given quantity of water, we reduce the impurity more than if we wash once with double the quantity. For, let the original impurity be 1 gram, and let us add 10 grams of wash-water and filter off 10 grams : there will then remain $\frac{1}{10}$th of the original impurity. At the second washing there will remain $\frac{1}{10}$th of that, or $\frac{1}{100}$ of the original. If we had added the 20 grams at once, the impurity would have been only reduced to $\frac{1}{21}$. It is evident that for the same amount of wash-water we shall get the best result by using small quantities at a time, and washing many times.

The following table gives an approximation to the smallest volume of wash-water, and minimum number of washings required, to reduce the precipitate to a given state of purity. It is obtained by regarding the apparent volume of the precipitate at the bottom of the beaker or on the filter as consisting *wholly* of a solution of impure matter, which it is required to reduce to a certain degree of purity. by successive dilutions with a constant volume of water.

Let v be the volume of the precipitate at the bottom of the beaker or on the filter, regarded as above, v the amount of wash-water used at each washing, n the number of washings. Also let $\frac{1}{a}$ be the fraction of the original amount of impurity which remains after n washings, then

$$\left(\frac{v}{v+\mathrm{v}}\right)^n = \frac{1}{a} \quad \cdot \quad \cdot \quad \cdot \quad \cdot \quad (1)$$

Further, if w be the total volume of water employed in the n washings, $\mathrm{w} = n\,\mathrm{v}$, and (1) becomes

$$\left(1 + \frac{1}{n} \cdot \frac{\mathrm{w}}{v}\right)^n = a \quad \cdot \quad \cdot \quad \cdot \quad (2)$$

If we make n infinite, a well known algebraical theorem gives

$$\mathrm{W} = v \log_e a \quad \cdot \quad \cdot \quad \cdot \quad \cdot \quad (3)$$

and this value of W is the smallest volume of water by the use of which the impurity can be reduced to $\frac{1}{a}$ of its original amount.

$\frac{1}{100,000}$			$\frac{1}{50,000}$			$\frac{1}{0,000}$			$\frac{1}{10,000}$		
I.	II.	III.	I.	II.	III.	I.	II.	III.	I.	II.	III.
$\frac{v}{v}$	n	W	$\frac{v}{v}$	n	W	$\frac{v}{v}$	n	W	$\frac{v}{v}$	n	W
0·5	28·4	14·2	0·5	26·7	13·3	0·5	24·4	12·2	0·5	22·7	11·4
1	16·6	16·6	1	15·6	15·6	1	14·3	14·3	1	13·3	13·3
2	10·5	21·0	2	9·8	19·7	2	9·0	18·0	2	8·4	16·8
3	8·3	24·9	3	7·8	23·4	3	7·1	21·4	3	6·6	19·9
4	7·1	28·6	4	6·7	26·9	4	6·1	24·6	4	5·7	22·9
5	6·4	32·1	5	6·0	30·2	5	5·8	27·6	5	5·1	25·7
6	5·9	35·5	6	5·6	33·4	6	5·1	30·5	6	4·7	28·4
7	5·5	38·8	7	5·2	36·4	7	4·8	33·3	7	4·4	31·0
8	5·2	42·0	8	4·9	39·4	8	4·5	36·1	8	4·2	33·5
9	5·0	45·0	9	4·7	42·3	9	4·3	38·7	9	4·0	36·0
10	4·8	48·0	10	4·5	45·1	10	4·1	41·3	10	3·8	38·4
11	4·6	51·0	11	4·4	47·9	11	4·0	43·8	11	3·7	40·8
12	4·5	53·9	12	4·2	50·6	12	3·9	46·3	12	3·6	43·1
13	4·4	56·4	13	4·1	53·3	13	3·8	48·8	13	3·5	45·4
14	4·2	59·4	14	4·0	55·8	14	3·7	51·1	14	3·4	47·5
15	4·2	62·3	15	3·9	58·5	15	3·6	53·6	15	3·3	49·8
16	4·1	65·0	16	3·8	61·1	16	3·5	56·0	16	3·3	53·0
17	4·0	67·8	17	3·7	63·6	17	3·4	58·3	17	3·2	54·2
18	3·9	70·4	18	3·7	66·1	18	3·4	60·5	18	3·1	56·3
19	3·8	74·3	19	3·6	68·6	19	3·3	62·8	19	3·1	58·4

By taking the logarithm of formula (1) we obtain

$$n = \frac{\log a}{\log\left(1 + \dfrac{V}{v}\right)} \quad . \quad . \quad . \quad (4)$$

the logarithms being common logarithms, and this formula enables us to find the least number of washings requisite to bring down the impurity to a fraction $\dfrac{1}{a}$ of its original amount, by the use of a quantity v at each washing. The foregoing table has been calculated from it. The topmost line of the heading shows the fraction $\dfrac{1}{a}$.

By employing for each treatment the same volume of wash-water, and approximately determining the relative volumes of the precipitate, and of the washing liquid, used each time, we may obtain from the table on the preceding page, calculated from the foregoing formula, the minimum number of treatments required to reduce the impurity in the precipitate to $\frac{1}{100000}$, $\frac{1}{50000}$, $\frac{1}{20000}$ or $\frac{1}{10000}$ of its weight. Column I. gives the relation of the volume of the precipitate to the volume of the washing-fluid employed for each treatment. Column II., the minimum number of treatments necessary for the particular extent of washing desired, and Column III., the total volume of wash-water which will be obtained. (See p. 57.)

Let us suppose that we have a precipitate occupying a volume at the bottom of the beaker of thirty cubic centimetres, and that the amount of liquid we employ to wash it each time is fifteen cubic centimetres, then $\dfrac{V}{v}$ is of course 0·5, and if we wished to remove the impurities to the $\frac{1}{50000}$th part, we learn from Column II. of the table that we must treat it *at least* 27 (26·7) times with this amount of water (viz., 15 c.c.)— that is to say, the *minimum* amount of wash-water needed will be 399 cubic centimetres. In cases of simple decantation from beakers, the volume occupied by the precipitate, as compared with the fluid above it, may be very easily determined by laying a strip of paper along the side of the beaker, marking off the height of the precipitate, and level of the liquid, and finding the number of times the length corresponding to the height of the precipitate may be folded into the length corresponding to the depth of the supernatant liquid. In applying this table to filters, the capacity c of these must be calculated; it is given by the formula

$$c = \frac{\pi r^3}{8\sqrt{3}}$$

where r is the radius of the filter-paper.

The following table shows the capacity in cubic centimetres of various filters, placed in a funnel whose opposite sides form an angle of 60°.

| No. 3 | 6·1 c.c. | No. 5 | 28·3 c.c. | No. 7 | 77·8 c.c. |
| ,, 4 | 14·5 ,, | ,, 6 | 49·0 ,, | ,, 8 | 116·1 ,, |

When the whole of the precipitate has been brought on to the filter, the unoccupied volume of the latter is determined by filling it with water from a burette. If w be the amount of water required to fill it, then $\frac{w}{c-w}$ gives the entry $\frac{v}{v}$ in Col. I. of the table on p. 57, whence we obtain the minimum number of washings required.[*]

The rapidity with which a liquid filters is proportional to the difference of pressure exerted on its upper and lower surfaces. By the ordinary method of filtration this difference seldom exceeds six millimetres of mercury. By increasing the difference we accelerate one of the most tedious of the operations of quantitative analysis. The following arrangement effects this acceleration to the desired extent. A glass funnel is chosen of about 8 centimetres in depth, the sides of which are free from irregularities or bulgings, and subtend an angle of 60°. The stem should be long and

---

[*] Bunsen, Ann. der Chem. u. Pharm., vol. cxlviii. p. 269. In the absence of exact knowledge respecting the nature of precipitates, whether pervious or impervious to liquids, and in what degree, or whether different liquids have different powers of adhering to or penetrating precipitates, we must regard the above process as an attempt only to place the operation of washing upon a quantitative basis.

narrow, and the end should be cut obliquely. A small cone, 1 to 1½ centimetre in depth, of thin platinum foil or gauze, and having exactly the angle of the funnel, is dropped into the apex, and over it is fitted the filter, with all the precautions described on p. 53. The following is the readiest method of obtaining the platinum cone of the desired shape. A circular piece of writing paper, 10 or 12 centimetres in diameter, is folded like a filter, and placed in the funnel so as to fit accurately to its sides, especially near its apex. It is kept in position by a few drops of sealing wax, and is saturated with oil by means of a feather, care being taken that no *drops* of the oil remain at the point of the paper cone. A thin cream of plaster of Paris is then poured into the paper mould, and a small handle is inserted into it just before the mass becomes solid. In a few hours the plaster cast will be dry enough to be removed from the funnel, together with the oiled paper. The outside of the paper is now thoroughly oiled, and inserted into a small crucible, or similarly shaped vessel, of 4 or 5 centimetres in height, filled with cream of plaster of Paris. As soon as the outer mould is dry, the plaster cone is removed, and the paper rubbed off it. In this manner a solid cone, fitting into a hollow cone, is obtained, both of which possess exactly the angle of

FIG. 25.

FIG. 26.

inclination of the funnel (fig. 25). A piece of platinum foil, of such thickness that 1 square centimetre weighs 0·15 gram, is cut into the shape and size represented in fig. 26 : it is divided by a pair of scissors along the line *a b*, as far as the centre *a*. The foil is then held in the flame of the lamp for a few minutes, to render it pliable, and placed against the plaster cone, so that the point *a* is at the end of the cone ; the side *a b d* is folded against the cone, and over this is folded the

remainder, *a b c,* so that the foil also becomes a cone, the
sides of which have exactly the same inclination as those of
the plaster cast, and also of the funnel. The shape of the
platinum funnel may be completed by dropping it into the

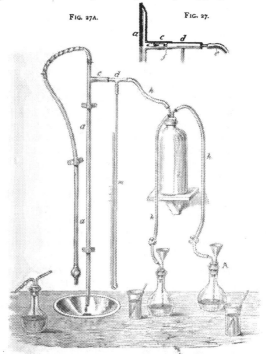

FIG. 27A.   FIG. 27.

hollow mould, and pressing it down by means of the plaster
cone: this shape of course may, at any time, be again given to it
by simple pressure between the two cones. The platinum cone
should allow no light to pass through its apex: when properly made it will support a filter filled with liquid, under a
pressure of an atmosphere, without the paper breaking: the

small space between the folds of the foil is quite sufficient to
allow of the passage of a rapid stream of water from the filter.

The stem of the funnel is pushed through a caoutchouc
cork, pierced with two holes, and fitting into a *thick* glass
flask (A, fig. 27); the second hole carries a piece of glass
tube ending immediately under the cork and leading to the
instrument which creates the difference in pressure. This
is also seen in fig. 27. A brass tube, *a a*, about 1
metre in length and 8 millimetres in internal diameter, has
its upper end cut obliquely in the manner seen in fig. 27A.
At about 5 centimetres from the end is a side tube *c* of
equal diameter and 5 centimetres long, into which is screwed
a short piece of tube *d* ; the ends of this tube *d* are fitted
with narrow brass tubes, *e* and *f*, 4 millimetres in diameter
and 2 centimetres long. Over *f* is pushed a piece of thick
caoutchouc tube 4 centimetres in length. This tube must
be made of *good* caoutchouc : it should be about 6 millimetres in external diameter, and its bore should not exceed
2 millimetres in width. Before introducing it into *c*, a piece
of wood, somewhat wider than its bore, is pushed into it,
and the caoutchouc is cut through to the wood by a smart
blow on the head of a chisel, 2 centimetres broad, placed
against the tube at 15 millimetres from the end. The wood
is then withdrawn, and the end of the caoutchouc tube is
stopped air-tight by a short length (1 centimetre) of glass
rod, held firmly in position by binding wire. The thick
caoutchouc tube so cut, forms a very efficient valve, which,
on the application of pressure from within, opens, but closes
immediately by outward pressure. The tube being of considerable thickness in the walls, is rigid, and does not collapse
even under a pressure of an extra atmosphere. The upper
end of the tube *a a* is connected by means of a short piece
of elastic caoutchouc tubing with the water-supply ; this
tube should be bound round with calico to within 5 or 6
centimetres of the end near the brass tube, since it will be
subject to considerable inward pressure. On allowing a
sufficient amount of water to flow through, it commences
to pulsate as the india-rubber valve intermittently opens and

shuts. Rapid suction is thus set up, and the instrument exhausts a closed vessel in a comparatively short time to within the pressure due to the tension of aqueous vapour corresponding to the temperature of the water flowing down the tube, *plus* the tension required to open the caoutchouc valve. The degree of exhaustion is determined from the height of the mercury in the manometer *m*, which is connected with the tube *d* by means of a piece of strong caoutchouc tubing. The entire apparatus is fixed upon a board, which may have a foot if it is desired to move it from place to place in the laboratory ; or it may be fixed in a position where the water can most conveniently flow away.* By connecting the tube *h* with the flask holding the funnel (or with an intermediate vessel to which several flasks are attached) we diminish the pressure to which the under surface of the liquid to be filtered is exposed, so that the filtrate is driven with greatly increased rapidity through the pores of the paper ; the filter itself is prevented from being pushed through into the stem by the closely-fitting little platinum cone which supports it.

The diminution of pressure may also be readily brought about by the aid of the little apparatus seen in fig. 27B, which is specially applicable to water under high pressure. The apparatus is attached by a stout piece of caoutchouc tubing to the water tap; the water flowing in in the direction indicated by the arrows. When forced through the narrow internal tube *b* into the sharply bent fall-tube *c c*, a partial vacuum is

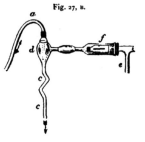

Fig. 27, b.

created in the bulb-shaped portion, *d*, and hence within the

* For an explanation of the principle of this apparatus, see a paper by Mendelejeff, Kirpitschoff, and Schmidt, Ann. der Chem. u. Pharm., January 1873. See also Jagn, Ann. der Chem. u. Pharm., Feb. 1873.

filter-flask or other vessel with which it is connected by
the T-tube *e*, which may also be put in connection, if necessary, with a manometer.   In the tube *f* is a small caoutchouc valve, similar to that described in the apparatus shown
in fig. 27A, to prevent the possible reflux of the water.

To use this apparatus for filtering, the liquid resting over
a precipitate is cautiously poured on to the filter fitted to the
funnel, with the precautions detailed on p. 53, the action of the
pump is set up, and as the liquid flows through into the flask
fresh portions are added until the whole has been decanted.
The precipitate is then transferred in the ordinary manner and
washed by the addition of water from *an open-mouthed vessel*,
and not by a jet from the wash-bottle.   The fluid in which
the precipitate was originally formed, together with that
necessary to transfer the precipitate to the filter, should be
allowed to flow away completely before the process of washing is commenced.   Immediately after the precipitate is
drained, *but before any channels commence to form in it*, the
filter is to be filled up with water, poured cautiously down
the side of the funnel.   When this wash-water has drained
away, the suction is continued until the precipitate is seen
to shrink, when the filter is again filled up *over the edge and
to within* 1 *centimetre of the brim of the funnel.*   This process
is to be twice repeated, after which the precipitate may be
drained almost dry by continuing the action of the pump
for a few minutes.   This method of filtration and washing
is exceedingly rapid as compared with the old plan, and
requires very little wash-water by reason of the compression
which the precipitate suffers.   Thus a precipitate of chromium
sesquioxide weighing about 0·24 gram, required 1 hour 48
minutes, and 1050 cubic centimetres of water to wash it to
within $\frac{1}{80000}$ by the old method, whilst with the new plan
the same amount of sesquioxide required only from 12 to 14
minutes, and from 39 to 41 cubic centimetres of water.
(Bunsen.)

A further advantage attending the use of the suction
apparatus arises from the condition of the precipitates after

filtration. The chromium sesquioxide, for example, is left so dry, that without further desiccation, the precipitate, wrapped in the filter, may be placed in the crucible over the lamp, and, after cautiously charring the paper, may be ignited without any apprehension of loss by projection. Many other precipitates which experience no alteration when ignited in contact with paper, such as ferric oxide, alumina, &c., may be treated in the same way. The paper being nearly dry may also be readily detached from the funnel ; when opened out on a flat surface, the coherent precipitate may easily be removed by means of a thick platinum wire, so as scarcely to leave a trace upon the filter. This ready method of removing the precipitate is of great value when we have occasion to treat it with a solvent or flux.

This suction apparatus may be used for a variety of purposes in addition to that of filtration ; it may be employed as an aspirator in quantitative operations, since the volume of air passing through the tube can be regulated with the utmost nicety by the aid of the screw-clamp ; it may also be applied to the evacuation of desiccators or vessels in which the concentration of liquids *in vacuo* is conducted, to freeing crystals from mother-liquors, &c.

Drying the precipitate.—In the majority of cases it is necessary to dry the precipitate thoroughly before it can be further treated with the view of determining its weight. The water in the stem of the funnel is removed by filter-paper, and the mouth of the funnel closed, to protect the precipitate from dust, by placing a moistened filter over it ; on drying, the paper adheres to the rim and makes a very efficient cover. The funnel is then placed in the steam-bath represented in fig. 13 (p. 38), and kept there until the paper and precipitate are completely dried. This method of drying the precipitate is preferable to that of supporting the funnel directly over the lamp, for in addition to the risk of cracking the stem, the latter method has the further disadvantage of causing the precipitate, by reason of the manner of heating, to adhere

F

to the paper. When dried in the steam-bath, the precipitate, in contracting, detaches itself from the filter ; so much so, that many curdy or gelatinous precipitates like silver chloride, or ferric or chromic oxides, may be almost completely shaken out of the funnel into the crucible in which they are to be weighed. This ready separation of the dried precipitate from the paper materially conduces to accuracy in determining its weight.

Igniting and weighing the precipitate.—Since the precipitate requires to be weighed in a perfectly dry state, it is in general necessary to remove it from the filter and to ignite it. A porcelain or platinum crucible is heated, allowed to cool in the desiccator, and weighed. It is then placed on the sheet of black glazed paper, together with the platinum wire and feather (Fig. 21, p. 51). The filter is removed from the funnel, opened out, and the detached fragments of the precipitate allowed to fall into the crucible. The portions of the precipitate adhering to the filter are loosened by rubbing its sides together, care being taken that its surface is not thereby destroyed, otherwise filaments of paper are apt to contaminate the precipitate ; these may either escape burning in the subsequent ignition, or if burnt, may alter the composition of the precipitate. The detached precipitate is then added to the main quantity already in the crucible. Care must be taken to remove as much of the precipitate as possible from the filter ; but however carefully the operation may be performed, a considerable amount of the substance invariably remains, either on the surface of the paper or contained within its pores. When the substance suffers no change by ignition with carbonaceous matter, it may be recovered by burning the paper, and adding the ash to the crucible. The known weight of the filter-ash is then subtracted from the increase in the weight of the crucible. The filter is burnt in the manner described on p. 51. The paper should be so folded that the soiled half of the filter is

in the centre : there is thus less chance of loss from projection
or from the precipitate fusing to the heated platinum wire.
The ash is shaken into the crucible, which is then ignited, at
first gently, and with the lid on ; afterwards more strongly, and
with the lid removed.   The degree and duration of the heat
depend, of course, on the nature and amount of the precipitate : as a general rule from five to ten minutes at a low
red heat will be sufficient.   The crucible is then placed in
the desiccator, and weighed when cold.   It must be heated
a second time and again weighed, to ascertain that its
weight is constant.

Some precipitates suffer change when ignited in contact
with carbonaceous matter, or become altered in composition
at the high temperature necessary to burn the carbon completely.   Thus silver chloride becomes reduced to metallic
silver in contact with carbon, and calcium carbonate is
converted into caustic lime at a red heat.   In weighing
silver chloride, for example, the precipitate is detached as
far as practicable from the filter, and the crucible in which
it is placed is gently heated until the chloride fuses, and
when cold it is weighed. The paper is now folded in the
usual way, the soiled portion being in the centre, and it is
burned in the manner described, and the ash added to the
fused silver chloride.   The crucible is again weighed : its
increase of weight gives the amount of the filter-ash, together
with the quantity of *metallic silver* which has been reduced
from the state of silver chloride by contact with ignited carbonaceous matter.   Since 108 parts of silver correspond to
143·5 of silver chloride, the amount of silver chloride in the
pores of the paper can be readily calculated from this
reduced silver : it is of course added to the weight of the
main quantity of the chloride.   The cases in which it is
necessary to weigh the filter-ash separately will be mentioned
as they occur.

Whenever practicable, a platinum crucible should be employed on account of the readiness with which it may be

heated to redness. Indeed, in some cases, its use is almost
indispensable, as in the conversion of calcium oxalate into
carbonate, and of magnesium-ammonium-phosphate into
magnesium pyrophosphate. Platinum vessels, however,
cannot be used for the ignition of silver chloride or bromide

FIG. 28.

or of lead chloride. Many oxides,
sulphides, and phosphides cannot
be heated in contact with platinum without injury to the metal.
After prolonged ignition over a
lamp, especially if the reducing
portion of the flame be permitted
to impinge upon it, the lower portion of the crucible loses its lustre
and appears to be superficially corroded. This appearance
is said to be due to the formation of a carbide of platinum.
Red-hot platinum crucibles should never be touched with
brass tongs or placed in brass rings, as black stains are thus
formed on the metal. They are best heated on pipe-clay
triangles or on thin platinum triangles supported on a triangle
of stout iron wire, fig. 28. Clean iron tongs will be found
more generally convenient than brass tongs. Platinum
crucibles may be cleaned by scouring with moist *sea-sand,*
which does not scratch the metal; stains which cannot thus
be removed are got rid of by heating with acid potassium
sulphate, or borax, the crucible being afterwards thoroughly
washed with hot water and scoured with sea-sand.

Collection of precipitates on weighed filters.—Occasionally
we have to deal with a precipitate which cannot be ignited to
ensure the expulsion of moisture before being weighed. The
precipitate must be weighed therefore on the filter on which
it is collected. Accordingly the weight of the filter itself
must be known. The paper is folded in the usual manner,
placed in a stoppered tube, or between well-fitting watch-
glasses pressed together by a clip, and heated in the
steam-chamber for an hour or so. The stoppered tube,

or watch-glasses, together with the filter-paper, are allowed to cool in the desiccator, and weighed when cold. The filter is then fitted into the funnel, and the precipitate is brought on to it, the tube or the watch-glasses being meanwhile left in the balance-case. The paper and precipitate are first dried in the funnel, the filter is then detached from the glass, and placed in the tube or between the watch-glasses, heated for some hours in the bath, and repeatedly weighed until the weight is constant.

Another plan is to weigh two filters of equal size (A and B) against each other, and mark the difference in weight on B. The precipitate is collected on A, the filtrate and washings being allowed to pass through B; both are dried and weighed against each other, and the original difference in weight allowed for.

PART II.

SIMPLE GRAVIMETRIC ANALYSIS.

I. COPPER SULPHATE. $CuSO_4 + 5H_2O$.

Preparation.—In order to obtain this salt in a fit state for analysis, it is necessary to purify it by recrystallisation. The blue vitriol of commerce not unfrequently contains ferrous sulphate, which cannot be removed even by repeated crystallisation, as the two sulphates tend to crystallise together. By heating the solution with a few drops of nitric acid, the ferrous salt is oxidised to ferric sulphate, and on concentrating the liquid, crystals of pure copper sulphate are easily obtained. Two hundred grams of clean, well-developed crystals of the commercial salt are dissolved in about a quarter of a litre of hot water, a few drops of nitric acid are added, the solution is filtered, if necessary, and boiled for half-an-hour ; on cooling the liquid deposits crystals of the pure salt. After standing for a few hours the solution is poured off, and the mother-liquor is drained as far as possible from the crystals. The crystalline mass is broken up by means of a glass rod, and dried by pressure between folds of filter-paper. It is advisable not to use too great a degree of force in pressing the salt, as the sulphate may thus become mixed with filaments of filter-paper, which interfere with the accuracy of the analytical operations. When the greater portion of the moisture has been removed by repeated pressure between filter-paper, the salt is wrapped in a fresh sheet of dry paper, and the folds are placed under a heavy weight for an hour or two. Whilst the salt is drying, the apparatus required for its analysis is got ready. Two small thin test-tubes, to hold about 6 or 8 grams of the salt, are cleaned and dried, and fitted with good, clean, soft corks. A couple of beakers of 250 cubic centimetres capacity, and two large watch-glasses to cover their mouths, a filter-flask fitted with its funnel, and two thin glass rods, all perfectly clean and dry, are also needed.

The complete analysis of copper sulphate necessitates the determination (1) of the water of combination ; (2) of the copper ; and (3) of the sulphuric acid.

1. *Determination of the water of combination.*—Copper sulphate gives up the whole of its combined water on heating ; 4 molecules being readily expelled at 100-110°, and the remaining molecule at about 200°. The determination of the amount of water expelled at different temperatures may be made by means of the air-bath (fig. 14, p. 39), or in the apparatus represented in fig. 29. The test-tube (*a*) contains the tube and salt to be dried; it is about 8 centimetres long, and 2 centimetres wide ; into it is placed the narrower and shorter tube, containing the weighed amount of salt : the tube *a* is closed with a cork pierced with two holes, into which are fitted narrow glass tubes bent at right angles : one of these tubes passes nearly to the bottom of the test-tube. The narrow tubes are connected by means of caoutchouc tubing with the small flasks *c* and *d*, containing strong sulphuric acid : the tube *e* of the flask *d* is in connection with the filter-pump, by means of which a current of air, dried by aspiration through the acid in the flask *c*, is drawn over the salt. The test-tube dips into a small quantity of oil contained in a beaker of 400 cubic centimetres capacity. The tube is held by means of a clamp attached to a retort-stand. The oil is heated by a small flame, and the temperature is ascertained by a thermometer placed near to the tube.

By the time this piece of apparatus has been fitted up, and the beakers, &c., are cleaned, the copper sulphate under the weight will be dry. One of the test-tubes which has been fitted with a cork is nearly filled with the dried salt ; any adhering powder is wiped from the upper portion and edge of the tube, and the cork is replaced. The remainder of the copper sulphate is set aside in a stoppered bottle ; it will be useful for subsequent analytical operations.

Weigh the other test-tube and cork, and introduce about

Fig. 29.

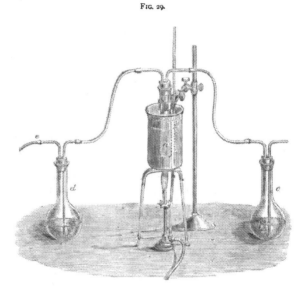

1·5—2 grams of the copper salt, taking care not to soil the edge of the tube ; replace the cork and weigh again. The increase of weight gives the amount of salt taken for the determination. Take out the cork, and leave it in the balance case. Drop the little tube containing the salt into the wider test-tube of the drying apparatus, insert the cork and bent tubes, heat the oil-bath to 100-110°, with frequent stirring, and aspirate a gentle current of air through the sulphuric acid. In about an hour the greater portion of the water will have been expelled: the little tube is withdrawn from the wider one by means of the forceps, allowed to cool in the desiccator, and carried to the balance-room. When cold, the cork in the balance-case is inserted into the tube, and the loss of water which the sulphate has suffered is determined by a second weighing. The cork is once more withdrawn, left in the balance-case, and the tube again heated in the oil-bath to 100-110°, a current of dry air being swept over it for about half-an-hour, after which it is again weighed when cold, in order to determine if it has parted with an additional quantity of water. If the second weighing is within 0·0010 gram of the first, the loss may be set down as constant; but if the weighings differ by more than this amount, the tube and salt must be again heated for half-an-hour, and weighed a third time, the process being repeated until a constant weight is obtained.

The temperature of the oil-bath is next raised to 200°, and maintained at this point for about an hour, a current of dry air being passed uninterruptedly through the apparatus. After cooling, the loss of weight experienced by the salt (which is now nearly white) is again determined, the tube is once more replaced in the bath, again heated, and again weighed, the operation being repeated until no further loss of water is perceptible.

Arrange the results of the several weighings in the following manner in your note-book. The numbers here given are the results of an actual determination.

ANALYSIS OF COPPER SULPHATE.

(Date.)   1. *Determination of Water.*

Tube + cork + salt	6·3400 grams.
Tube + cork	4·8905 ,,
Weight of salt taken	1·4495

Weight of tube + cork + salt after drying :

After 1 hour at 105°-112°		5·9340	grams
After 30 min. additional : 110°		5·9210	,,
,,	,, 107°-110°	5·9180	,,
,,	,, 107°-112°	5·9176	,,
Water expelled at 110°		0·4224	,,
or 29·14 per cent.			
After drying for 1 hour at 190°-200°		5·8250	,,
After further drying for half an hour at 200°		5·8183	,,
,, ,, ,, 205°		5·8176	,,
Additional loss of water		0·1000	,,
or 6·91 per cent.			

2. *Determination of the Copper.*—Whilst the salt is drying in the oil- or air-bath, proceed with the estimation of the copper. This is effected by precipitating it as cupric oxide, by the addition of caustic soda solution.

$$CuSO_4 + 2NaHO = CuO + Na_2SO_4 + H_2O.$$

The corked tube containing the copper sulphate is carefully weighed, and about a gram of the salt is shaken out into one of the clean and dry beakers. On replacing the cork, and again weighing the tube, its loss of weight gives the exact amount taken for analysis. Care must, of course, be taken that all the copper sulphate removed from the tube finds its way into the beaker. The salt is then dissolved in about 50 cubic centimetres of hot water : the solution should be perfectly clear, and free from suspended matter. It is boiled, the mouth of the beaker being meanwhile closed by one of the large watch-glasses, in order to prevent the projection of any of the solution on ebullition. The lamp is drawn aside, and a clear solution of caustic soda is poured into the liquid, drop by drop, down the side of the beaker, the liquid meanwhile being kept in agitation. A precipitate is at once formed ; it is at first of a bluish green colour, but it rapidly darkens as it falls through the hot liquid, and ultimately becomes nearly black. These changes of colour are due to the passage of the copper oxide from the hydrated to the anhydrous condition. The precipitate is allowed to settle, when, if sufficient soda has been added, the liquid will be colourless. Ascertain that the alkali is in excess by testing the solution with a slip of reddened litmus-paper. Fold a No. 5 filter, drop it into the platinum cone, moisten thoroughly with hot water, and fit it carefully to the funnel in the manner already described (see p. 53). Next slightly grease the edge of the beaker, and by means of the glass rod decant the clear liquid on to the filter (taking care not to disturb the precipitate of copper oxide) and set the pump in operation. When the whole of the liquid has been decanted on to the filter, pour about 30 or 40 cubic centimetres of hot water over the precipitated oxide, boil for a few minutes with the glass cover on, allow to settle, and again pour the supernatant liquid on the filter. Repeat the washing by decantation and then carefully rinse every particle of the precipitate with hot water on to the filter.

It may happen that a small quantity of copper oxide obstinately adheres to the beaker and cannot be removed by washing. Pour about 2 cubic centimetres of hot water into the beaker, and a couple of drops of nitric acid : by the aid of the glass rod, the dilute acid solution may be made to dissolve the adhering oxide. This is reprecipitated by the addition of a few drops of soda solution, and thrown on to the filter. Now fill up the filter five times with hot water, taking care to allow the whole of the wash-water to run through before a fresh addition is made. If these instructions are followed the precipitate will be thoroughly washed. The funnel is withdrawn from the flask, its mouth is covered with paper to protect the copper oxide from dust, and the whole is placed in the drying chamber. Whilst the precipitate is drying, clean, dry and heat a small porcelain crucible and lid (No. 1 size), place them together in the desiccator, and when cold, carefully weigh them. When the copper oxide is dry, it is detached *as far as possible* from the filter, and transferred to the weighed crucible. The funnel is cleaned, if necessary, by rubbing it with the edge of the paper ;

the filter is burned, and the ash allowed to fall into the crucible. *One drop* of nitric acid is allowed to moisten the oxide and filter-ash, and the crucible is gently warmed until the mass is dry, when the heat is raised until the bottom of the crucible is red hot. It is allowed to cool slightly, and whilst still warm, is transferred, together with its lid, to the desiccator, and when quite cold, again weighed. The crucible and lid are once more heated, and again weighed on cooling ; care should be taken that the reducing gases from the flame do not find their way into the crucible. The second weighing ought not to differ more than 0·5 milligram from the first weight. If the difference is greater, the operation must be repeated until a constant weight is obtained. The increase in weight of the crucible and lid gives the amount of copper oxide contained in the quantity of salt taken for analysis *plus* the ash of No. 5 filter. The details of an actual determination will show how the results ought to appear in the note-book :

ANALYSIS OF COPPER SULPHATE.

2. *Determination of Copper by precipitation with Caustic Soda.*

(1)	Weight of tube + salt	. .	10·6052
(2)	,, ,,	. .	9·5805
	Salt taken	. .	1·0247

Weight of crucible + lid + CuO + ash (1)	8·1530
,, ,, ,, (2)	8·1527
Crucible + lid . . .	7·8240
	0·3287
Less filter-ash No. 5	0·0023
	0·3264

Calculation :

$$\frac{63·1 \times 0·3264 \times 100}{79·1 \times 1·0247}$$

log 63·1	1·8000
log ·3264	1·5137
log 100	2·0000
	3·3137
log 79·1	1·8982
	1·4155
log 1·0247	·0105
	1·4050 = 25·41 Cu.

3. *Determination of the Sulphuric acid.*—This is effected by adding barium chloride to the solution of the copper salt and weighing the precipitated barium sulphate :

$$CuSO_4 + BaCl_2 = BaSO_4 + CuCl_2$$

About one gram of the copper sulphate is weighed out into the second beaker, and is dissolved in 30-40 cubic centimetres of water, a few drops of hydrochloric acid are added, and the solution is heated to the boiling point. The lamp is now drawn aside and solution of barium chloride is added drop by drop. In order to determine whether an excess of the precipitant has been added, the barium sulphate is allowed to subside, and when the liquid is sufficiently clear, a drop of the barium chloride solution is poured down the side of the beaker. If an increased turbidity ensues, the liquid is again heated and a further quantity of barium chloride added : the precipitate is once more allowed to settle, and the liquid again tested by the cautious addition of barium chloride. When you have assured yourself that the precipitation is complete, cover the beaker and set it aside in a warm place for a few hours. If you attempt to filter the turbid liquid immediately, the finely divided precipitate will inevitably pass through the pores of the paper, but on standing, especially after precipitation from a hot solution slightly acidified with hydrochloric acid, the barium sulphate becomes denser and more granular. When the precipitate has completely settled add one more drop of barium chloride to again assure yourself that the precipitation is complete, and proceed to collect the barium sulphate. Fit a No. 5 filter carefully into the funnel, grease the edge of the beaker, pour the clear liquid, by the help of the glass rod, on to the filter, and cautiously set the pump working. When the whole of the liquid has passed through, pour 40-50 cubic centimetres of hot water over the barium sulphate and boil the liquid for a few minutes ; the precipitate is allowed to settle and the liquid (which will now be slightly turbid) poured on to the filter. It is well to stop the flow of water in the pump until the first filter-full of liquid has passed through, otherwise there is danger of the precipitate making its way through the filter. When the whole of the liquid on the filter has passed through, fill up the funnel again with the wash-water, cautiously set the pump in operation, rinse the precipitate on to the filter with hot water, remove any barium sulphate adhering to the beaker by means of a feather, and wash five or six times with hot water. Take care to allow the whole of the liquid to pass through before a fresh quantity is poured in, or the filtrate may become turbid. Dry the precipitate and transfer it from the filter to a weighed porcelain crucible, burn the filter and add the ash to the crucible. Place the crucible on the triangle and *cautiously* heat it with the lid on for 2 or 3 minutes, increase the heat and keep the bottom red hot for 5 or 10 minutes, allow to cool in the desiccator, and when cold, weigh : repeat the heating and again weigh. The two weighings ought to agree. The results should thus appear in the note-book :

ANALYSIS OF COPPER SULPHATE.

3. *Determination of Sulphuric Acid by precipitation as Barium Sulphate.*

Weight of tube + salt (before)	9·5803
,, ,, (after)	8·6115
	0·9688

(1) Crucible + lid + BaSO$_4$ + ash	8·7317
(2) ,, ,, ,, ,,	8·7315
Crucible + lid	7·8242
	0·9073
Less ash No. 5	·0023
	0·9050

Calculation : $\dfrac{96 \times 0\cdot9050 \times 100}{233\cdot1 \times 0\cdot9688}$

log 96	1·9823
log ·905	1·9566
log 100	2·0000
	3·9389
log 233·1	2·3676
	1·5713
log ·9688	1·9863
$SO_4 =$	1·5850 = 38·46 p. ct.

The complete analysis is therefore as follows :

	Calculated			Found
Cu		63·1	25·33	25·41
SO₄		96·0	38·54	38·46
4H₂O expelled at 100°–110°		72·0	28·90	29·14
H₂O ,, 200°		18·0	7·23	6·91
		249·1	100·00	99·92

II. SODIUM CHLORIDE (Na Cl).

Preparation.—Common salt rarely contains more than 98 per cent. of sodium chloride, its principal impurities being calcium sulphate and magnesium chloride. These sub-stances cannot be readily removed by recrystallisation, but by adding hydrochloric acid to a strong solution of the salt, pure chloride of sodium is precipitated, and the magnesium chloride and calcium sulphate remain dis-solved. About 70 grams of salt are dissolved in a quarter of a litre of hot water, the solution is filtered and saturated with hydrochloric acid gas. The apparatus represented in fig. 30 may be conveniently used for this purpose. The flask *a* contains the salt and sulphuric acid, and the evolved hydrochloric acid gas, after passing through a small quantity of the acid solution contained in the bottle *b*, is led into the filtered brine. The exit-tube of the apparatus is replaced by a small funnel dipping into the solution of salt : this method of delivering the gas into the liquid prevents the possibility of the precipitated sodium chloride interfering with the pas-

sage of the gas by closing the outlet.   The salt begins to
separate out almost immediately, and in an hour or so the
process may be interrupted.   The liquid is poured from the
precipitated salt, which is washed once or twice with pure
strong hydrochloric acid solution, allowed to drain, and
heated gently in a porcelain basin.   The moisture cannot

FIG. 30.

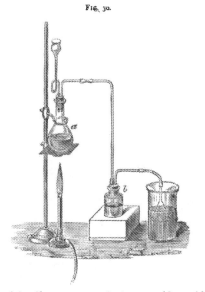

be removed by filter-paper, as the strong acid would cause
the contamination of the salt with the iron, &c., contained
in the ash.   The mass should be heated gently in a porcelain
crucible until all the acid is expelled, powdered roughly
while still warm, and a portion introduced into a small dry
tube fitted with a good cork.   The remainder of the salt is
placed in a stoppered bottle : it will prove useful in subsequent operations.

1. *Determination of the Chlorine.*—This is effected by pre-cipitation as silver chloride, by means of silver nitrate solution.

$$NaCl + AgNO_3 = AgCl + NaNO_3.$$

About 0·5 gram of the salt is weighed out into a beaker of 80 cubic centimetres capacity and dissolved in 30-40 cubic centi-metres of water; a few drops of pure nitric acid are added, together with a solution of silver nitrate. If sufficient silver solution has been added, the chloride separates out as a dense curdy precipitate. When you have satisfied yourself that all the chlorine is precipitated, heat the liquid to near the boiling point, stirring it occasionally by means of a glass rod, and allow the precipitate to settle by placing the beaker, protected from the dust, in a warm place for a few hours. Be careful to protect the silver chloride as much as possible from the light. Pour the clear liquid on to a filter, wash twice by decantation with hot water, carefully rinse the precipitate on to the filter, wash it 5 or 6 times with hot water and dry it in the steam bath. Transfer the dried chloride, detached as completely as possible from the filter, to a weighed porcelain crucible and heat very gradually, increasing the temperature until the chloride begins to fuse at the edges of the mass; allow to cool in the desiccator and weigh. If the chloride has been carefully protected from light it will have at most but a slight violet tinge, and the fused portion will have the appearance of horn. The filter-paper is folded in the manner described on p. 67, and burnt, the ash being allowed to fall on to the chloride. The crucible and its contents are again weighed : the increase in weight gives the amount of metallic silver originally adhering as silver chloride to the filter, together with the ash of the paper. The known weight of ash in the filter subtracted from the total increase gives the amount of reduced silver; this is calculated to silver chloride, and the amount added to the main quantity. An actual example may make this clearer :

G

$$\text{Sodium Chloride taken} \quad 0\cdot4065 \text{ grams}$$

Crucible + AgCl		8·9710	,,
,, ,, + Ag + ash		8·9813	,,
	Crucible	7·9860	,,

$$8\cdot9813 - 8\cdot9710 = 0\cdot0103$$
$$\text{Less ash} \quad 0023$$
$$\overline{\text{Ag} = 0080} = \text{AgCl} \quad 0\cdot0106$$

$$8\cdot9710 - 7\cdot9860 = 0\cdot9850$$
$$\text{Total AgCl} = 0\cdot9956$$

$$\text{Cl} = \frac{0\cdot9956 \times 35\cdot5 \times 100}{143\cdot5 \times 0\cdot4065} = 60\cdot60 \text{ per cent.}$$

The fused silver chloride may readily be detached from the crucible by placing over it a small piece of zinc and adding a few drops of dilute sulphuric acid. The semi-reduced mass, together with the silver chloride, precipitated by adding a few drops of hydrochloric acid to the filtrate should be put into a bottle labelled ' silver residues.' When these residues have sufficiently accumulated they are to be worked up as directed in the Appendix.

Determination of the Sodium.—The salt is converted into sodium sulphate by the action of strong sulphuric acid. Clean, ignite, and weigh a platinum crucible and lid, introduce into it 0·5 to 1 gram of salt, and again weigh. The increase of weight gives the amount of sodium chloride taken. Place the crucible on a triangle in the slanting position represented in fig. 15, and add drop by drop pure strong sulphuric acid. Do not heat the crucible for ten or fifteen minutes, or until the reaction has moderated. There is no danger of any of the substance being lost by projection, if care is taken to place the crucible as directed, and to add the acid cautiously. Now heat the crucible gently from the top, placing the lid as indicated in the figure, and allow the flame to approach the bottom of the crucible very gradually. The operation must be done slowly, and with constant watching.

or a portion of the sulphate may be lost by spirting. In a few minutes the whole of the hydrochloric acid will have been expelled, and dense fumes of sulphuric acid will be evolved. As these diminish, the heat is gradually raised, until the crucible attains a full red heat. Maintain it at this temperature for fifteen or twenty minutes, put on the lid, allow to cool, and weigh. Again heat to redness for five minutes, and weigh a second time. The operation is to be repeated, until a constant weight is obtained. The fused mass should be quite white.

III. PEARL-ASH.

Good pearl-ash of commerce contains from 93 to 96 per cent. of potassium carbonate, the rest consisting of water, alkaline sulphates, chlorides, silica, &c. To determine its value it is merely necessary to estimate the percentage amount of potassium carbonate and water. The quantity required to make up 100 is taken as the measure of the impurities.

Determination of the Moisture.—Weigh out from 3 to 4 grams of the coarsely-powdered ash from a tube into a small weighed porcelain crucible provided with a lid. Gently heat the ash for half-an-hour over a small gas flame, and weigh when cold; again heat, and again weigh. It must not be forgotten that potassium carbonate is highly hygroscopic; the weighings must, therefore, be made as expeditiously as possible.

Determination of the Potassium.—The carbonate is converted into the double chloride of platinum and potassium ($PtCl_4 2KCl$). Weigh out about 2 grams of the carefully sampled carbonate and dissolve in 30–40 c.c. of water. Filter, if necessary, into a 250 c.c. flask, wash the filter thoroughly from every trace of alkali, fill the flask up to the mark with distilled water, and shake vigorously. Transfer two lots of 50 c.c. each to porcelain basins, cover the basins with glass plates, and add dilute solution of hydrochloric acid

in slight excess. Heat the basins on the water-bath, and
when the expulsion of the carbonic acid is finished, rinse
the covers with a few drops of hot water, and add solution of pure platinum tetrachloride. It is necessary to
add the solution of platinum in considerable excess; about
1 gram of the metal is required for 0·5 gram of the carbonate. Evaporate the solutions on the water-bath until
they become pasty ; remove the dishes, and, when nearly cold,
pour over the crystals about 25 cubic centimetres of rectified
methylated spirit. Do not attempt to break up the highly
crystalline scales of the double salt. The action of the
alcohol in dissolving out the excess of platinum tetrachloride
may be facilitated by imparting a gentle rotatory motion to
the contents of the dish.* Cover, and allow to stand for
five or ten minutes ; pour the clear liquid on to a No. 2
filter (which ought to be first washed with hot water, and
then with alcohol), and drain the liquid as far as possible
from the precipitate : again add about 10 cubic centimetres
of spirit to the precipitate, and shake ; allow to stand five
minutes, and pour the clear liquid through the filter.
Repeat the digestion with spirit a third and fourth time : the
solution will now be nearly colourless. Transfer the precipitate to a weighed crucible, by the aid of a glass rod, and a
stream of alcohol from a small wash-bottle ; pour the alcohol
through the filter, and wash the paper carefully with alcohol
until the filtrate is absolutely colourless. Dry the double
chloride at 70°, heat to 100° in the steam-chamber, and
weigh the precipitate. If the above instructions have been
properly followed, scarcely a stain will be left on the paper.
The filter is burnt, the ash added, and the whole again
weighed : the increase in the weight of the crucible after
subtracting the ash, may without sensible error be considered
as platinum : it is calculated to K_2PtCl_6, and the amount

* Water would dissolve the double chloride : 100 parts of water
dissolve about 1 part of the salt at the ordinary temperature ; whereas
the same quantity of the salt requires 26,400 parts of alcohol of 80 per
cent. and 42,600 parts of absolute alcohol.

added to the main quantity. When calculated to potassium carbonate, the result of the two determinations ought not to differ by more than 0·2 per cent.

IV. ROCHELLE SALT. $C_4H_4KNaO_6 . 4H_2O$.
(*Separation of Potassium and Sodium.*)

Preparation.—40 grams of cream of tartar, and 30 grams of crystallised sodium carbonate, are added successively in small portions at a time, to 150 cubic centimetres of boiling water. The liquid must be tested to ascertain that it is alkaline ; it is filtered, if necessary, and concentrated by evaporation. On cooling, the solution deposits fine large crystals of the potassium-sodium-tartrate.

Analysis.—Powder a few of the crystals, press between filter-paper, and weigh off from a tube about 1·5 gram into a platinum dish or crucible. Heat gently, so as to melt the salt in its water of crystallisation, and gradually increase the flame until the mass is perfectly dry. Ignite at a low red heat in the draught-chamber for some time ; allow the charred mass to cool, and digest it repeatedly with hot water ; filter, acidulate with pure hydrochloric acid, and evaporate to dryness in a weighed platinum dish, with all the precautions necessary to avoid loss by spirting. As soon as the residue is perfectly dry, heat it very gently for five minutes over the lamp, transfer to the desiccator, and when cold weigh the mixed chlorides. Dissolve in a small quantity of water, transfer to a porcelain dish, and separate the potassium with platinum tetrachloride, as directed in the foregoing example. Enough platinum chloride must be added to convert both the alkaline chlorides into the double salts of platinum. The sodium-platinum-chloride is readily soluble in alcohol, especially if previously moistened with water.

V. DOLOMITE.

This substance is essentially a double carbonate of lime and magnesia. No definite relation, however, exists between the amounts of the two carbonates, as the calcium and magnesium replace each other in all proportions. Occasionally the mineral occurs associated with the isomorphous carbonates of iron and manganese.

The portion employed for analysis is finely powdered, dried in the steam-bath, and introduced into a small corked tube.

Determination of the Carbonic Acid.—The mineral is decomposed by dilute hydrochloric acid, and the carbon

FIG. 31.

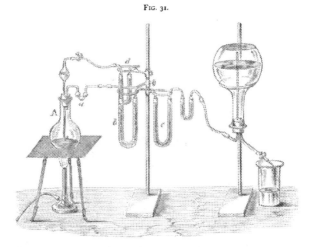

dioxide, freed from moisture, is absorbed by soda-lime. The determination is most accurately effected by means of the apparatus represented in fig. 31. The flask A has a capacity of about 150 cubic centimetres : it is fitted with a caoutchouc cork, containing two holes, into one of which is inserted the little bulb-tube *a*, containing a few drops of strong sulphuric acid. This serves to regulate the rapidity of the decomposition, by indicating the speed with which the gas travels through the apparatus. The tube *b*, which

may be 10 centimetres high, is filled to a depth of 6 centimetres with pumice saturated with solution of copper sulphate, and strongly heated until all the water has been
expelled : its use is to absorb any vapour of hydrochloric
acid which may pass over from the flask. The remainder
of the tube is filled with coarsely-powdered calcium chloride. In the tube *c* the carbonic acid is absorbed ; seven-
eighths of it are filled with soda-lime ; and as this substance, in combining with carbonic acid, becomes heated,
and parts with a little water, the remaining one-eighth of
the tube is filled with calcium chloride : *c* is also connected with a small unweighed tube containing calcium
chloride, in order to prevent the possibility of the weighed
tube absorbing atmospheric moisture. Caoutchouc corks are
used to close the tubes ; if, in their absence, ordinary corks
are employed, they must be cut, and covered with sealing-
wax. The tubes may be suspended as in the figure, or by
wires from a glass rod running through a cork held in a
clamp on the retort stand. When not in use the tubes are
closed by stoppers of glass rod. In the other hole of the
caoutchouc stopper of A is fitted a bulb-tube, passing down
nearly to the bottom of the flask, drawn out and turned
up in the manner represented in the figure ; this can be
closed by means of a small screw clamp : *d* contains soda-
lime : it is attached to the caoutchouc tubing on the bulb-
tube at the close of the experiment.

Weigh out about 1·5 gram of the carbonate into A, and
add about 10 cubic centimetres of water. Weigh *c* without
the stoppers and tubing, and put the several parts of the
apparatus together, setting the flask A upon a piece of wire
gauze on a tripod. Nearly fill the bulb-tube with hydrochloric acid solution, diluted with its own volume of water.
Open the clamp, and allow enough acid to pass into the
flask to set up the evolution of the carbonic acid, and as the
action diminishes add more acid, until the tube is empty.
Close the clamp, and connect the exit-end of the apparatus
with an inverted wash-bottle, of about 600 cubic centimetres

capacity, and filled with water, the jet of which you have transferred to the short, obtusely-bent tube, and temporarily closed by a piece of caoutchouc tubing and clamp; heat A slowly to boiling, boil for about one minute, cautiously open the clamps, remove the lamp, and aspirate a slow current of air through the apparatus, regulating the speed by means of the screw. Detach *c*, allow it to cool, and weigh it; the increase of weight gives the amount of carbonic acid in the mineral. This method of estimating carbon dioxide is very accurate, and is generally applicable; it is expeditious, and has the advantage of being direct. The tubes may be used a great number of times without their contents being changed, if they are well stoppered when not in use. A very simple and accurate method of determining carbon dioxide in salts and minerals consists in heating the substance with fused and powdered potassium bichromate in a short combustion-tube (about 25 cm. long), passing the evolved gas through a tube containing calcium chloride, and absorbing it in a weighed soda-lime tube. Or a weighed portion of the carbonate is heated with about four times the quantity of fused borax in a platinum crucible to dull redness, and the carbon dioxide determined from the loss of weight.

Determination of the Silica, Iron (and Manganese), Lime, and Magnesia.—Decant the solution from the flask A into a porcelain basin, rinse the flask with a few cubic centimetres of hot water, adding the washings to the main quantity of the liquid, and evaporate the whole to complete dryness on the water-bath, in order to render the small quantity of silica insoluble. Moisten the dried residue with a few drops of strong hydrochloric acid, cover the basin with a glass plate, allow to stand for a few minutes, add hot water, and filter the solution through a No. 3 filter; wash thoroughly, drain the paper by the action of the pump, fold the filter, and without further drying throw it into a weighed platinum crucible : cautiously heat until the paper is dry, and incinerate. The increase in weight of the crucible, *minus* the filter-ash, gives the amount of silica.

To the filtrate from the silica, add a little bromine-water, then ammonium chloride and ammonia in slight excess ; heat gently for some time and filter. Wash the precipitate once or twice, re-dissolve it in hydrochloric acid by heating, add one more drop of bromine, again precipitate with ammonia, and filter. Dry and weigh the oxide of iron (and manganese *): the precipitate may be ignited without complete drying. The second precipitation effects the removal of small quantities of lime and magnesia precipitated with the ferric oxide.

Mix the ammoniacal filtrates, and add ammonium oxalate in quantity sufficient to precipitate the lime and to convert the magnesia into oxalate. Presence of excess of ammonium oxalate prevents the slight solubility of calcium oxalate in chloride of ammonium solution. Allow the liquid to stand for ten or twelve hours. Decant the clear liquid on to a filter and wash once or twice by decantation, taking care to disturb the precipitate as little as possible. Dissolve the calcium oxalate in a small quantity of hydrochloric acid, heat to boiling, add a few drops of ammonium oxalate and a slight excess of ammonia. This double precipitation of the calcium oxalate effects the separation of a small quantity of co-precipitated magnesia. Filter off the oxalate, wash thoroughly with hot water, and dry. Transfer the dried precipitate to a weighed platinum crucible, and if its weight does not exceed 1 gram, proceed to convert it into caustic lime. Burn the filter, and add the ash to the crucible. Heat the crucible gently, with the lid on, and gradually increase the flame until the bottom is red hot. Now expose the crucible to a full red heat over the blow-pipe for fifteen minutes, occasionally removing the lid for a few seconds ; allow to cool in the desiccator, and weigh. By this treatment the oxalate is converted, first into carbonate and then into caustic lime. The

---

* For a method of determining the manganese, usually present in small quantity only, in limestones and dolomites, see Part IV.

heating must be repeated until the weight is perfectly constant; that is, until the carbonate is wholly converted into lime. If the quantity of the oxalate exceeds 1 gram, its conversion into oxide is accomplished with difficulty and requires prolonged heating. In this case it is better to transform the oxalate simply into carbonate by heating gently over a small flame scarcely sufficient to make the bottom of the crucible appear red hot by diffuse daylight. The conversion into carbonate is rendered visible by a slight change of colour which creeps over the heated oxalate. After heating for ten minutes, weigh, and repeat the operation with the lid on until the weight is constant. Weigh the filter-ash and the small quantity of adhering lime on the lid. Moisten the carbonate with a few drops of water and test it with a slip of reddened litmus paper; it ought not to show the slightest trace of alkalinity. If the paper becomes blue, the crucible has been overheated; in that case transfer the ash of the filter to the crucible, add a few drops of ammonium carbonate solution, evaporate to dryness on the water-bath, heat very gently for a few minutes and weigh. The ammonium carbonate reconverts the caustic lime into carbonate. The filtrate from the oxalate of lime is poured into a porcelain basin, concentrated considerably, and rendered strongly acid by nitric acid. About 3 grams of the acid are employed for each gram of the sal-ammoniac supposed to be present. The dish is covered with an inverted funnel and gently heated, when a rapid effervescence sets in, owing to the decomposition of the ammonium chloride. The liquid is evaporated to dryness, when, if sufficient nitric acid has been added, all the sal-ammoniac will have been expelled.* The funnel is rinsed, and the saline mass in the dish dissolved in water, ammonium chloride added in small quantity, the solution rendered strongly alkaline by ammonia, filtered if necessary, and mixed with solution of sodium phosphate. The liquid

* This method of removing the excess of sal-ammoniac is preferable to that of evaporating to dryness in a platinum basin and igniting. By the latter plan there is danger of loss from the tendency of the salt to creep over the side of the dish during the evaporation.

is well agitated by shaking the beaker, covered with a glass plate, and set aside for twenty-four hours. The clear liquid is poured on to a filter and the precipitate washed by decantation in the beaker, and afterwards on the filter by dilute ammonia-water (1 part strong ammonia and 5 of water), until the filtrate acidulated with pure nitric acid gives only a slight opalescence with silver nitrate. Pure water dissolves the precipitate to a slight extent (1 part in 15,300 of water); in dilute ammonia it is much less soluble (1 part in 45,000). The presence of a large excess of ammonium chloride increases its solubility; hence the necessity of expelling the greater portion of this salt before precipitating the magnesia. The dried phosphate is detached as completely as possible from the filter, transferred to a weighed platinum crucible and very gradually heated (at first with the lid on) for fifteen minutes. The temperature is now raised until the crucible is red hot, when the lamp is withdrawn and the filter-ash added. Take care that the filter is burnt as completely as possible, or the crucible will be corroded by the action of the carbon on the phosphate. The crucible is then strongly ignited for a quarter of an hour, allowed to cool, and weighed. The residue (magnesium pyrophosphate $Mg_2 P_2O_7$) should be white, or, at most, have a very slight tinge of grey.*

VI. Barium Sulphate ($Ba SO_4$).

The pure substance is prepared by adding a dilute solution of barium chloride to an excess of hot and moderately diluted sulphuric acid. The precipitate is washed once or twice with hot water, dried, and ignited. The method of analysis is founded upon the decomposition of the sulphate by prolonged fusion with an alkaline carbonate.

About 1 gram of the ignited precipitate is weighed out

* In any analysis involving the separation of several substances it is advisable to preserve the weighed precipitates in small corked tubes or between watch-glasses until the analysis is finished. Questions regarding the purity or identity of the substances separated frequently arise, which, of course, cannot be answered if the bodies are thrown away.

into a platinum crucible and mixed with 3 parts of a dry mixture, in equivalent proportions, of pure potassium and sodium carbonates. This mixture may be conveniently obtained in a state of purity by igniting Rochelle salt $(C_4H_4KNaO_6+4H_2O)$ in a platinum basin, extracting the charred residue with hot water, filtering and evaporating to dryness. The mixture of barium sulphate and alkaline carbonates is fused at a bright red heat for thirty or forty minutes, allowed to cool, the mass extracted with water containing a few drops of ammonia and ammonium carbonate, and filtered.* The filtrate contains the sulphuric acid in union with the alkalies; the residue consists of barium carbonate. It is washed with water containing a few drops of ammonia and ammonium carbonate, dried, and weighed. It is dissolved in the crucible in a few drops of hydrochloric acid, and a slight excess of sulphuric acid added; the mixture is cautiously evaporated to dryness, and the residue ignited and weighed. Its amount ought to be equal to that originally taken.

The filtrate containing the alkaline sulphate is acidulated with hydrochloric acid, heated to boiling, barium chloride added, and the precipitate washed, dried, ignited, and weighed. Its weight ought to equal that of the barium sulphate analysed.

The same process is applicable to the analysis of strontium and calcium sulphates.

VII. INDIRECT ESTIMATION OF BARIUM AND CALCIUM.

It is frequently possible to determine the amount of a substance A, by combining it with a second body B, so as to form a definite compound A B. By estimating the quantity of B in the combination, we can readily calculate the amount of A which must be present. Thus we can determine the amount of Ag in a solution by estimating the amount of Cl required to precipitate it completely. In like manner we

---

* Barium carbonate dissolves in 14,000 parts of cold water and 15,500 of boiling water. It is ten times less soluble in water containing a slight quantity of ammonia and ammonium carbonate.

could determine the amount of Ba and Ca from the quantity of CO_2 respectively contained in their carbonates. Such indirect determinations are based *on the law of constant proportion*, which states that *the same substance always consists of the same elements united in the same proportion.*

But it will be obvious on a little reflection that we can determine the amount of barium and calcium in a mixture of their carbonates, by estimating the amount of carbon dioxide contained in a known weight of the mixed compounds. The possibility of such an estimation is based upon the wide difference which exists between the combining weight of barium (137·2) and that of calcium (40·0), both of which substances combine with 44 of CO_2. Supposing that we had found that 2 grams of the mixed carbonates had evolved 0·67 gram of carbon dioxide. Then, if the whole of the carbon dioxide were combined with the barium, the amount of the barium carbonate would be 3·002 grams.

Eq. CO_2.		Eq. $BaCO_3$.		CO_2 found.		
44	:	197·2	::	0·67	:	= 3·002.

But the weight of the mixed carbonates taken was only 2 grams. The deficiency (3·002 − 2·0)=1·002, is proportional to the amount of calcium carbonate present. This amount is thus found : *The difference between the atomic weights of* $BaCO_3$ *and* $CaCO_3$ *is to the atomic weight of* $CaCO_3$, *as the difference found is to the quantity of calcium carbonate contained in the mixture.*

$$BaCO_3 - CaCO_3 = 197·2 - 100 = 97·2$$

$BaCO_3 - CaCO_3$.		$CaCO_3$.				
97·2	:	100	::	1·002	:	= 1·032.

Accordingly the composition of the mixture is

Calcium Carbonate	1·032
Barium Carbonate	0·968
	2·000

or expressed centesimally,

Calcium Carbonate	51·6
Barium Carbonate	48·4
	100·0

Similarly we might determine the proportion of the bases present in the mixture by estimating the weight of sulphuric acid necessary to form the two sulphates. The method of calculation is, *mutatis mutandis*, precisely similar to that above given.

Weigh out into a platinum crucible about equal weights of pure and recently ignited barium and calcium sulphates (0·5 gram of each is a convenient quantity to take). Mix with 3 or 4 pts. of the mixture of sodium and potassium carbonates, and proceed exactly as described in the preceding example. The weighed barium and calcium carbonates are then decomposed in the apparatus described in p. 84, and the amount of carbon dioxide determined with great care. From the weight of CO_2 obtained, the proportion of the two bases is calculated in the manner above described.

As a control, determine the amount of sulphuric acid in the filtrate, after acidulation with hydrochloric acid, by precipitation with barium chloride, according to the method described on p. 75, and again calculate the proportion of the two bases. This exercise will afford a good trial of the manipulative skill of the operator.

VIII. Ferrous Ammonium Sulphate.
$$Fe(NH_4)_2 \, 2SO_4 + 6H_2O.$$

To prepare this salt, 27·8 grams of recrystallised ferrous sulphate and 13·2 grams of pure ammonium sulphate are separately dissolved in the least possible quantity of water at a temperature of about 40°. The solutions are mixed, a few drops of sulphuric acid are added, and the mixture is stirred constantly until cold. The greater portion of the salt separates out in a finely-divided state : if the solution is now set aside for a few hours a further quantity of the double salt crystallises out. Pour off the mother-liquor, allow the crystalline powder to drain, and dry it thoroughly between filter-paper.

Determination of the Ammonia.—Weigh out about 1 gram

of the salt into the retort (fig. 32), the neck of which is contracted at *a*, and the upper portion filled with fragments of broken glass. The tube *b* is filled with strong soda-lye, which can be delivered little by little on opening the clamp. The flask *c* is fitted with a caoutchouc cork and bent tube *e*, on which is blown a bulb. The short wide tube *d* is filled with fragments of glass, previously well washed with water ; through this tube hydrochloric acid is poured. The tube *e* is so arranged that it just dips beneath the surface of

Fig. 32.

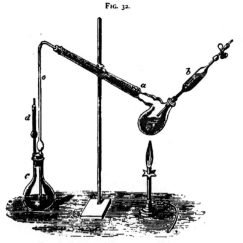

the liquid in the flask. The retort containing the weighed quantity of ammoniacal salt, dissolved in a small quantity of water, is fixed on a clamp, the tube *e* inserted into its neck, a small quantity of soda solution allowed to flow into the retort, and the liquid heated to boiling. Care must be taken to prevent the liquid, if it shows any tendency to froth, from passing over into the flask. The broken glass in the neck of the retort tends to prevent such a mishap. The liquid should be boiled for fifteen or twenty minutes after the caustic soda solution has been added. When the evolution of ammonia is finished, the tube *e* is disconnected, the powdered glass in *d* washed with distilled water, the tube *e* drawn up from the liquid and washed with distilled water. The ammoniacal solution is poured into a porcelain basin, an excess of platinum tetrachloride is added, and the whole is evaporated *just* to dryness on the water-bath. The double chloride is washed with strong alcohol, and is transferred to the weighed platinum crucible in the manner described on p. 82. The salt is dried at 70° or 80°, and heated to 100° for ten or fifteen minutes, after which its weight will be constant. The little filter is dried and ignited on the lid of the crucible and weighed separately. The weight of the residual platinum is calculated to that of double salt, and the amount added to the main quantity. By way of control, the double salt may be *gently* heated so as to expel the greater portion of the ammonium chloride ; the crucible is then raised to a full red heat, and the metallic platinum weighed. Determining as platinum, however, is not more accurate than weighing the double salt, owing to the readiness with which finely-divided particles of the metal are carried away in the vapour of the escaping ammonium chloride.

IX. Determination of Nitric Acid.

Dr. Gladstone and Mr. Tribe have found that a thin plate of zinc coated with copper (formed by placing the former metal in a solution of copper sulphate for a few minutes) decomposes water, particularly on warming, with the formation of zinc hydrate and the evolution of hydrogen.

The hydrogen so eliminated is capable of reducing nitric acid in combination to the state of ammonia :

$$NO_3K + 4H_2 = NH_3 + HKO + 2H_2O.$$

This reaction constitutes the basis of a method of determining nitric acid in nitrates.

About 25-30 grams of thin sheet zinc are placed in a flask

of about 200 c.c. capacity, and covered with a moderately-concentrated and slightly-warmed solution of copper sulphate. In about ten minutes a thick spongy coating of copper will be deposited on the zinc : the liquid is poured off the metals, which are now well washed with cold water, and covered with about 40 or 50 c.c. of pure water. Weigh out about 0·5 gram of pure nitre into the flask, which is then

FIG. 33.

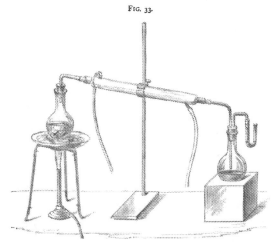

placed in a sand-bath and connected with the condensing arrangement shown in fig. 33. The receiver and U tube contain a few cubic centimetres of dilute hydrochloric acid. The liquid is gradually heated to boiling and distilled for about an hour. The distillate is poured from the receiver and evaporated to dryness in a porcelain basin over the water-bath, with excess of platinum tetrachloride, and the double chloride is treated exactly as in the foregoing example.

H

X.   POTASH-ALUM. $Al_2(SO_4)_3.K_2SO_4.24H_2O$.

The salt is purified by recrystallisation, powdered, dried between blotting-paper, and placed in a well-corked tube.

Determination of the water.—About 1 gram of the double salt is heated to 120° in the apparatus represented in fig. 29, until it ceases to lose weight. The loss is equivalent to 10 atoms of water. The temperature is now raised to 200°, and the heat maintained at this point until the weight is once more constant. The salt should now be perfectly anhydrous.

Determination of the Alumina and Potassium Sulphate.— Weigh out about 1 gram of the crystalline salt into a porcelain basin, dissolve in hot water, and add a moderate quantity of ammonium chloride solution, together with a *slight* excess of ammonia. Heat the liquid to boiling and maintain it in gentle ebullition for some time, keeping the basin covered with a sufficiently large watch-glass to avoid loss by projection. Rinse the watch-glass into the basin, and pour the clear supernatant liquid on to the filter, wash the precipitate once or twice by decantation and transfer it also to the filter ; wash three or four times with boiling water, and, after the whole of the liquid has passed through, keep up the action of the pump for ten minutes. Remove the filter containing the precipitate from the funnel, and, without further drying, place it in a weighed platinum crucible ; heat gently for a few minutes to char the paper, and gradually increase the flame until the crucible is red hot. Keep it at this temperature for ten or fifteen minutes, occasionally removing the lid, and then ignite it strongly over the blow-pipe for five or ten minutes ; place the crucible in the desiccator, and weigh when cold.

Ignition over the blast-lamp expels the last traces of water, together with the minute quantity of sulphuric acid which is precipitated with the alumina from a solution containing sulphates.

The filtrate from the precipitate of alumina contains the potassium sulphate : it is evaporated to dryness in a weighed platinum basin, gently heated to expel ammoniacal salts, and moistened with a few drops of pure sulphuric acid, in order to transform the potassium chloride, which invariably forms when potassium sulphate is heated with ammonium chloride, back into sulphate. The mass is again heated, with all the precautions detailed on p. 82, in order to expel the excess of sulphuric acid, and when cold, the potassium sulphate is weighed.

XI. GLASS

consists of a mixture of the alkaline silicate with certain insoluble silicates, generally of calcium, lead, iron, aluminium, magnesium, or manganese. The best window glass has approximately the composition, $Na_2OCaO.6SiO_2$. In flint glass the lime is replaced by oxide of lead. The pale green variety used for chemical apparatus is mainly made up of silicates of lime and potash, mixed with smaller quantities of iron and alumina.

In order to analyse it, the glass is reduced to the finest possible state of division, and fused in a platinum crucible with four times its weight of a mixture of equal parts of sodium and potassium carbonates. When cold, the crucible is placed in a porcelain basin, and the mass boiled out with water, hydrochloric acid is added in excess, and the whole is evaporated to *complete* dryness over the water-bath. The dried mass is then moistened with strong hydrochloric acid, hot water is added, and the silica is filtered off, repeatedly washed with hot water, dried and weighed. The solution contains the lead, iron, alumina, manganese, lime, and magnesia. The alkalies cannot, of course, be determined in this portion, as they are mixed with the salts required to decompose the glass. Pass sulphuretted hydrogen through the filtrate, to precipitate the lead; filter, dry it, and convert it into sulphate by treatment with strong nitric acid. Add a few drops of bromine to the filtrate, and heat gently; add ammonia, and filter off the iron, alumina, and manganese. The lime and magnesia are separated as in No. V.

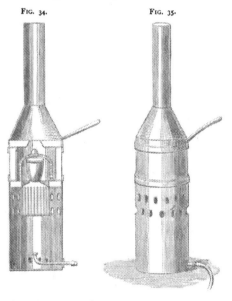

FIG. 34.　　　　　FIG. 35.

Determination of the Alkalies.—About 1·5 gram of the finely-powdered glass is weighed out into a platinum crucible, and intimately mixed with 9 grams of calcium carbonate, and 1·5 gram of ammonium chloride, and heated to bright redness for an hour, in a small table furnace (figs. 34, 35). The platinum crucible should be protected from the direct action of the fire by being placed in a small clay crucible, with a little calcined magnesia at the bottom. When cold the contents of the crucible are boiled with water in a silver or platinum dish, filtered, the filtrate acidified with hydrochloric acid, and evaporated to dryness to render the silica insoluble. The residue is treated with hot water and filtered; to the filtrate, ammonia, ammonium carbonate, and a few drops of ammonium oxalate are added to throw down the lime. The liquid is boiled, to render the precipitate dense and granular. It is filtered off; the liquid is evaporated to a small bulk in a porcelain basin, pure nitric acid is added in quantity, and the whole is evaporated to dryness to destroy the ammonium chloride. The saline residue is dissolved in a little water, and filtered if not quite clear, and again evaporated to dryness with a small quantity of strong hydrochloric acid, whereby the nitric acid is expelled. If the quantity of the mixed alkalies is considerable, this treatment with hydrochloric acid must be repeated once or twice before the nitric acid is completely dissipated. The alkaline chlorides are again dissolved in a little water, and evaporated to dryness in a weighed platinum dish, heated gently, and weighed. The potassium chloride is then separated by platinum tetrachloride in the manner described in No. IV. p. 83. Its amount, subtracted from the sum of the chlorides, gives the sodium chloride.

XII. FELSPAR (*Orthoclase, Albite*).

The group of the felspars may be regarded as silicates of alumina united, in varying proportions, with silicates of the alkalies and alkaline earths. The varieties, orthoclase and albite, differ from one another in crystalline form and in chemical composition. Orthoclase crystallises in forms belonging to the monoclinic system, and the alkali it contains is chiefly potash, whereas albite is triclinic, and its alkali consists mainly of soda.

The following analyses serve to show this characteristic difference in composition :—

	Orthoclase	Albite
Silica	64·76	67·62
Alumina	17·60	16·59
Ferric oxide	0·50	2·30
Lime	0·65	0·85
Magnesia	0·30	1·46
Potash	14·18	0·51
Soda	1·75	10·24
Loss on ignition	0·65	—
	100·39	99·57

The methods employed in the analysis of these minerals are identical with those described in No. XI. The only point which needs special mention is the separation of the iron and alumina.

The solution containing the iron, alumina, lime, magnesia, and alkalies, from which the silica has been removed by evaporation, is mixed with a little nitric acid, boiled for some time, and a *slight* excess of ammonia added, whereby, on boiling, the iron and alumina are precipitated. Care must be taken not to employ a very large excess of ammonia, otherwise the precipitation, even after protracted boiling, will not be complete. The mixed oxides are washed thoroughly with hot water, dried as far as possible by the action of the pump, and transferred to a platinum dish ; the small portions remaining on the filter are dissolved in hot hydrochloric acid, the solution being allowed to drop into the dish. When the whole of the acid solution has passed through, the dish is removed from under the funnel, a beaker put in its place, and the filter thoroughly washed with hot water, the washings being collected in the beaker. The precipitate in the dish is now dissolved in the least possible quantity of hydrochloric acid, an excess of a concentrated solution of pure caustic potash added, the liquid heated to boiling, and a lump of the pure hydrate added, in quantity sufficient to dissolve the alumina, and the boiling is continued for a few minutes. The contents of the dish are now washed into the beaker, diluted with a little water, and filtered, the ferric oxide being repeatedly washed with hot water, dried, and weighed. The alkaline solution containing the alumina is acidified with hydrochloric acid, a few crystals of potassium chlorate are added to destroy any organic matter present, which would tend to retain a small portion of the alumina in solution ; the liquid is concentrated in a porcelain basin, and ammonia is added in *slight* excess, and the liquid is boiled until a piece of turmeric paper held in the steam is no longer turned brown. The precipitate is filtered, dried, ignited over the blow-pipe in a platinum crucible, at first very gently, and with the lid on, and then for 5 or 10 minutes to bright redness.

Instead of treating the mixed oxides of iron and alumina with caustic potash, they may be washed, dried, ignited over the blow-pipe, and weighed together. The mixture, or an aliquot portion of it, is then brought into a porcelain boat, and strongly heated in a porcelain tube, in a current of dry hydrogen for an hour. The boat must be allowed to cool in the current of the hydrogen before it is withdrawn. The loss of weight which it suffers gives the amount of oxygen combined with the iron ; each milligram of loss is equivalent to 3·339 milligrams of ferric oxide. With proper care this method is very accurate. By way of control (and this is more particularly necessary when the amount of oxide of iron, compared with that of the alumina, is very small), you may treat the weighed mixture with highly dilute nitric acid (1 part of acid to 30 of water). The dissolved iron is re-precipitated, after filtering, by means of ammonia, dried, and weighed. The residual alumina is also dried and weighed.

XIII. Brass, Bronze, Gun-Metal, Bell-Metal.

(*Separation of Tin, Copper, Lead, Iron, and Zinc.*)

Weigh out about 5 grams of the finely-divided alloy (a portion of a penny, for example) into a flask, and dissolve in

25 c.c. strong nitric acid and 15 c.c. water at a gentle heat. Place a small funnel in the neck of the flask to prevent loss by spirting. When the substance is dissolved, add about three times the bulk of water, and digest the precipitate on the water bath with occasional shaking for about an hour. Allow the precipitate to settle, decant the clear liquid, and repeat the digestion on the water bath for about an hour with dilute nitric acid (1 : 6). The oxide of tin is thus obtained free from admixed metals ; it is filtered off, washed, dried, and weighed in a porcelain crucible. If the quantity is at all considerable, it requires to be ignited over the blow-pipe before it is rendered completely anhydrous. The filtrate is evaporated nearly to dryness with strong hydrochloric acid, to expel the greater portion of the nitric acid ; re-dissolved in hot water, and the solution precipitated by sulphuretted hydrogen. The clear liquid (which should smell strongly of the gas) is filtered off, and the precipitate washed once or twice by decantation with hydrochloric acid of sp. gr. 1·05, through which a stream of sulphuretted hydrogen is led, and afterwards with water containing sulphuretted hydrogen. The mixed sulphides are drained thoroughly, and transferred to a small porcelain basin, and digested with nitric acid and about 10 cubic centimetres of dilute sulphuric acid. The solution is evaporated nearly to dryness, a small quantity of water is added to dissolve the copper sulphate, and the lead sulphate is filtered off without delay through as small a filter as possible, and washed with water acidulated with sulphuric acid, the filtrate being received in a litre flask. When the copper has been washed away, the lead sulphate is washed with dilute alcohol to remove the last traces of acid, otherwise the filter-paper would blacken on drying and fall to pieces.* The sulphate is weighed in a porcelain crucible ; care must be taken to remove the precipitate as completely as possible from the

* The alcoholic washings are not to be mixed with the filtrate containing the copper.

filter before incinerating. The ash may be moistened with one drop of dilute nitric acid, heated gently, a drop of sulphuric acid added, and the contents of the crucible carefully dried and ignited. The filtrate in the litre flask containing the copper is diluted to the mark, and the liquid thoroughly mixed by shaking; 100 cubic centimetres are withdrawn, and the copper determined, as in No. I. Part. II. p. 72, by precipitation with soda. In a second portion of the solution determine the amount of metal by precipitation with metallic zinc. Transfer 800 c.c. to a porcelain basin, add an excess of pure sulphuric acid, and evaporate to dryness to expel nitric acid. Dissolve the copper sulphate in a small quantity of water, decant the solution into a weighed platinum dish, and place in it a piece of pure zinc (about 1 or 2 grams will be sufficient); add a few drops of hydrochloric acid, and cover the dish with a watch-glass. In about an hour the whole of the copper will be precipitated, partly as a coherent film on the dish, and partly in red, spongy masses. A drop or two of the supernatant fluid should be tested with sulphuretted hydrogen water; it should, of course, remain colourless. Assure yourself that the whole of the zinc is dissolved, press the spongy masses of copper together, decant the colourless liquid, and repeatedly wash the metal with boiling water until the washings give no opalescence when tested with silver nitrate or barium chloride. Allow the water to drain away, and cover the copper with a small quantity of strong alcohol. Pour this away, and dry the copper in the steambath. The precipitation of the copper may also be effected in a porcelain or glass dish; in this case the process requires longer time, owing to the absence of the galvanic action between the platinum and zinc. A weighed piece of platinum-foil placed in the dish, and of course weighed with it at the termination of the experiment, accelerates the operation.

The filtrates from the sulphides of lead and copper contain the iron and zinc; they are concentrated to a small bulk (about 70 cubic centimetres), filtered into a small flask, with a drop or two of nitric acid to oxidise the iron, heated for a few minutes, allowed to cool, and mixed with a small quantity of freshly precipitated barium carbonate suspended in water. The liquid should not contain too much free acid ; if a large excess is present, it must be removed by adding sodium carbonate before mixing with the barium carbonate. The flask is closed and occasionally shaken. After standing a few hours the iron is all precipitated. The liquid is filtered, ammonium chloride added, and the zinc precipitated by sulphuretted hydrogen. The zinc sulphide is filtered off, washed, re-dissolved in nitric acid, the solution boiled, and the zinc re-precipitated as carbonate by the addition of sodium carbonate. The zinc carbonate is filtered, washed, dried, and ignited in a porcelain crucible, and weighed as oxide.

The barium carbonate precipitate mixed with the iron is dissolved in hydrochloric acid, ammonium chloride is added, together with a slight excess of ammonia, and the liquid heated and filtered. The washed precipitate is re-dissolved in a few drops of hydrochloric acid, and the iron again precipitated by ammonia free from carbonate, washed, dried, ignited, and weighed as ferric oxide.

XIV.   GERMAN SILVER.

(*Separation of Copper, Zinc, and Nickel.*)

Weigh out about 1·5 gram of the finely-powdered alloy, and dissolve in nitric acid at a gentle heat, with the precautions mentioned in No. XIII. Evaporate the excess of acid, and separate any oxide of tin which may be formed. Precipitate the copper by sulphuretted hydrogen in hot solution: re-dissolve the copper sulphide, and again precipitate with sulpuretted hydrogen to separate the small quantity of co-precipitated zinc. The sulphide is then treated as in No. XIII., and the copper weighed as oxide.

To the filtrate containing the zinc and nickel is added a solution of sodium carbonate until a slight permanent precipitate is formed, which is then re-dissolved by the cautious addition of a few drops of hydrochloric acid. Pass sulphuretted hydrogen through the liquid, and to ensure the complete precipitation of the zinc add a few drops of a very dilute solution of sodium acetate, and again treat with sulphuretted hydrogen. Allow to stand for twelve hours, and filter off the zinc sulphide; wash it with sulphuretted hydrogen water, dissolve in hydrochloric acid, and precipitate with sodium carbonate, filter, wash, dry, and ignite and weigh as zinc oxide. Boil the filtrate containing the nickel after the addition of a few drops of hydrochloric acid, and precipitate with caustic soda (best in a porcelain basin), filter off the nickel hydrate, wash, dry, and ignite and weigh as nickel oxide.

XV. Britannia Metal.

(*Separation of Tin and Antimony.*)

About 1·5 gram of the alloy, as finely divided as possible, is oxidised in a porcelain basin with strong pure nitric acid, and evaporated to *perfect* dryness. The dried mass is then washed into a silver basin, again evaporated to dryness, and fused with an excess of sodium hydrate (about eight times the bulk). It is treated with a small quantity of water, and the liquid mixed with about one-third of its volume of strong alcohol. The stannate of soda, together with the excess of hydrate, is thus separated from the sodium antimoniate. The liquid is allowed to stand for six hours, filtered, and the precipitate washed first with weak spirit, and afterwards with strong alcohol. The antimoniate is dried, transferred to a porcelain crucible, and fused with potassium cyanide. Metallic antimony is thus obtained, which can be washed from adhering salts, dried, and weighed. The filtrate containing the tin is boiled to expel the alcohol, diluted if necessary, acidulated with dilute sulphuric acid, and precipitated by sulphuretted hydrogen. The tin sulphide is filtered off, washed, dried, and transferred to a weighed porcelain crucible, and cautiously roasted to oxide, and weighed.

XVI. TYPE-METAL.

(*Separation of Lead, Antimony, and Tin.*)

The alloy in fine powder is treated with nitric acid, and tartaric acid is added to the solution. The lead dissolves completely, together with the greater part of the antimony. The precipitated oxides are filtered off, and separated, as in No. XV. The solution containing the lead is evaporated to dryness with dilute sulphuric acid, and the lead sulphate separated as in No. XIII. The antimony and traces of lead in the filtrate are precipitated by sulphuretted hydrogen, the sulphide filtered off, washed into a flask, and mixed with an excess of yellow sodium or potassium sulphide. The flask should be closed with a good cork, and the solution kept at a gentle heat. Pour the clear liquid through a filter, and repeat the digestion with the alkaline sulphide twice. The residue consists of lead sulphide, which may also contain copper sulphide : these are separated as in No. XIII. Add hydrochloric acid to the alkaline filtrate, until the solution is distinctly acid ; allow the liquid to stand, and filter off the re-precipitated antimony sulphide. This is dried, transferred to a weighed porcelain crucible, moistened with strong nitric acid, and treated with ten times its weight of fuming nitric acid. The acid boiling at 86° must be employed for this purpose : nitric acid of sp. gr. 1·42 is not able to effect the complete oxidation of the sulphur, as its boiling point is about 16° higher than the fusing point of sulphur ; by heating with this acid the separated sulphur fuses, and forms little globules which resist oxidation. The white mass in the crucible consists of antimonic acid and sulphuric acid : by ignition it is converted into antimony tetroxide Sb$_2$ O$_4$. If the amount of sulphur mixed with the precipitated antimony sulphide is considerable, it is advisable, before proceeding to oxidise with nitric acid, to remove the greater portion of it by treatment with carbon bisulphide.

XVII. FUSIBLE METAL.

(*Separation of Bismuth, Lead, and Tin, with traces of Copper, Iron, and Zinc.*)

The alloy in the state of powder is oxidised with nitric acid, and the mass repeatedly digested (three or four times) with an excess of ammonia and yellow ammonium sulphide.

The mixture should be kept in a closed flask, and the solution gently warmed. In presence of copper, potassium sulphide must be used, as the sulphide of that metal is slightly soluble in ammonium sulphide. The tin is dissolved ; the bismuth and lead, together with the traces of copper, iron, and zinc, remain undissolved. The liquid is filtered, and the tin precipitated as sulphide by hydrochloric acid : it is filtered off, washed, dried, and roasted in a weighed porcelain crucible to the state of oxide. The sulphides of bismuth and lead, together with the small quantities of copper, iron, and zinc, are dissolved in nitric acid, evaporated nearly to dryness, water added, and the solution, without filtering, again treated with sulphuretted hydrogen. The lead and bismuth and traces of copper are thus once more precipitated as sulphides ; the zinc and iron remain in solution. The sulphides are next dissolved in nitric acid, sulphuric acid is added, the solution is evaporated to dryness, and the lead separated as sulphate. Nearly neutralise the solution by the cautious addition of ammonia, add a clear solution of common salt, and a large quantity of water.

Allow it to stand twenty-four hours, and test the supernatant liquid by adding a few drops of water : it ought to remain perfectly clear. The bismuth is thus completely precipitated as basic chloride (BiClO, or $BiCl_3 . Bi_2O_3$). It is filtered off, washed with cold water, dried, and fused in a capacious porcelain crucible with five times its weight of potassium cyanide. The fused mass is treated with water, when the metallic bismuth is left behind. The grains of the reduced metal are washed with water, and afterwards with spirits of wine, dried, and weighed. Separate the copper in the filtrate from the precipitated basic chloride of bismuth by sulphuretted hydrogen ; redissolve the copper sulphide, and precipitate as oxide by sodium hydrate. The iron and zinc are separated as in No. XIII.

PART III.

SIMPLE VOLUMETRIC ANALYSIS OF SOLIDS AND LIQUIDS.

WE have already indicated the principle of this mode of analysis, in showing how it is possible to determine the amount of silver in a liquid by the aid of a solution of hydrochloric acid, or of sodium chloride, of known strength ; and we have also shown how we can ascertain the amount of alkali in a solution of sodium or potassium hydrate by the use of litmus tincture, and of an acid solution of known chemical power. The following examples will serve to render the principles of volumetric analysis still clearer.

If we dissolve a small piece of iron-wire in dilute sulphuric acid, we obtain a solution of ferrous sulphate which is almost colourless, or which at most possesses a faint green tinge. If we add to the solution some substance which readily parts with its oxygen, the colour will change, the greenish tinge will give place to yellow, the ferrous salt becoming oxidised to the state of ferric oxide.

$$2FeO + O = Fe_2O_3.$$

A few grams of potassium permanganate ($KMnO_4$) dissolved in water give a deeply-coloured purple solution. Potassium permanganate, when in solution, very readily parts with its oxygen ; if we add a few drops of the liquid to the solution of ferrous sulphate, containing free sulphuric acid, we notice that the colour of the permanganate is instantly discharged. If, however, we continue to add successive quantities of the permanganate we arrive at a point when its colour is persistent. Let us consider what is the nature of this reaction. The potassium permanganate, in presence of free sulphuric acid, is decomposed ; permanganic acid is liberated, and potassium sulphate is formed. The permanganic acid, however, in presence of the ferrous sulphate and free sulphuric acid, readily parts with its oxygen, converting the ferrous salt into the state of ferric sulphate, and is itself reduced to the state of manganese sulphate, which in solution is colourless. So long as any ferrous sulphate remains in solution, this decolourising action will continue ; immediately, however, that the whole is converted into ferric sulphate, the red colour of the permanganic acid will remain unchanged.

This reaction may be represented by the equation :—

$$10FeSO_4 + 8SO_4H_2 + 2KMnO_4 = 5Fe_2(SO_4)_3 + K_2SO_4 + 2MnSO_4 + 8H_2O.$$

If we know the strength of the solution of permanganate—that is to say, if we determine the number of cubic centimetres we require to add to a solution containing a known weight of iron as ferrous sulphate, before the solution is permanently coloured—we can employ this solution of permanganate to determine the amount of iron in any given solution. Let us suppose that we required to add 50 cubic centimetres of permanganate solution to 0·5 gram of iron, dissolved in dilute sulphuric acid, before the colour was persistent; then each cubic centimetre of the permanganate would be equivalent to 0·01 gram of iron. If now we added the permanganate to a solution containing iron, say from an iron-ore, and found that we needed 25 cubic centimetres before the colour was permanent, we should say that the amount of iron in solution was 0·25 gram.

Instead of potassium permanganate, we may employ potassium bichromate as an oxidising agent. The reaction which occurs with this reagent may be thus represented :—

$$6FeSO_4 + K_2Cr_2O_7 + 7SO_4H_2 = 3Fe_2(SO_4)_3 + Cr_2(SO_4)_3 + K_2SO_4 + 7H_2O.$$

This equation tells us that 294·4 parts of potassium bichromate will convert 6 eq. or 336 parts of iron from the state of ferrous to that of ferric oxide. Potassium bichromate possesses a bright orange-red colour in solution, and if the products of its reaction on ferrous sulphate were colourless, we might continue to add the solution of bichromate until its colour was permanent, when we should know that, as the chromic acid was no longer decomposed, the whole of the ferrous salt was changed to the state of ferric salt. Unfortunately, however, the chromic sulphate $Cr_2(SO_4)_3$ which is produced has a deep green colour in solution which entirely masks the tint of the bichromate. We are accordingly obliged to have recourse to some other method than the persistency of the orange colour, to enable us to know when the whole of the ferrous oxide is converted into ferric oxide. Ferrous salts give a deep blue precipitate or colouration with a solution of ferricyanide of potassium ; ferric salts produce no such colouration. If then we sprinkle a few drops of the ferricyanide solution on a white surface, and from time to time take out a drop of the solution of iron which is undergoing oxidation, the gradual diminution in the intensity of the blue colour will inform us of the progress of the reaction, and its cessation will tell us when the oxidation is complete.

A consideration of these cases will enable us to lay down the conditions required in a volumetric process. In the first place, the reaction which constitutes the basis of the method must be constant, even under a diversity of circumstances. If, for example, it is modified by the concentration of the fluids, or the amount of free acid present, or if precipitates are formed during the reaction of variable composition, or if the presence of the air seriously affects the process, the reaction cannot, except in very special cases, afford the basis of a trustworthy method. A volumetric process further necessitates that we possess accurate means of determining the completion of the reaction. Thus the cessation of a precipitate in the case of the silver-salt, and standard solution of sodium chloride, denotes that the whole of the silver is precipitated. The change of the litmus tincture from blue to red indicates that the alkali is neutralised. The persistency in the colour of the permanganate solution tells us that the whole of the iron is in the state of ferric salt. In the case of the bichromate, we learn the same fact, from the non-formation of a blue colour with potassium ferricyanide. A *final reaction* must be sensitive, rapid, and decisive in its changes; if it requires considerable time, or a large expenditure of the testing fluid, or if it involves the passage through a series of closely-related tints or changes of colour, it cannot well serve to indicate the termination of the intended decomposition.

In order to carry out a volumetric process, we require :—

1. A solution of the reagent of known chemical strength : this we call a standard solution.

2. The means of accurately determining the completion of the reaction.

3. Accurate measuring vessels (pipettes, litre-flasks, &c.), and a graduated instrument termed a burette, for pouring

I

determinate quantities of the standard solution into the liquid on which it is to act.

The amount of apparatus specially required for volumetric analysis is not very extensive. In addition to a few beakers, flasks, porcelain basins, glass stirrers, &c., the student must

FIG. 36.

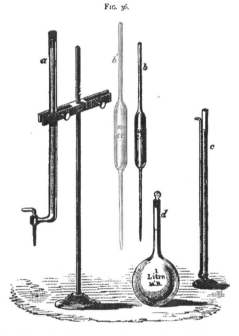

provide himself with a set of measuring flasks, pipettes, and burettes.

The most convenient series of measuring flasks is the following :—

(1) 1,000 cb.c.; (2) 500 c.c. ; (3) 300c.c.; (4) 250 c.c.; and

(5) 100 c.c. They should be fitted with well ground glass stoppers, and the graduation mark of each should be near the middle of the neck. The space between the mark and stopper allows the fluid to be more readily mixed by agitation. The flasks should be sufficiently thin to be heated without risk of fracture. Fig. 36 *d* represents a convenient form of litre-flask.

The following is the most convenient series of pipettes :— (1) 100 c.c. ; (2) 50 c.c.; (3) 25 c.c. ; (4) 10 c.c.; and (5) 5 c.c. Several 1 c.c. pipettes will be also needed ; these are readily made from glass tubing. The pipettes have the form seen in fig. 36 *b b*.

The measuring flasks and pipettes are generally sold with the graduating marks, their denomination, and the temperature at which the graduation is effected, etched upon them. But before employing them the operator must never neglect to verify their capacities. It must be borne in mind that pipettes are to be graduated to *deliver* their contents ; measuring flasks, however, should be graduated both to *contain* and to *deliver*. A 50 c.c. pipette, accordingly, needs to hold more than 50 c.c. of liquid ; it must hold this quantity *plus* that amount which adheres to the glass when the liquid is allowed to flow out. We frequently use the measuring flasks to dilute liquids to determinate volumes, from which we afterwards withdraw aliquot portions by means of the pipettes ; occasionally, however, it is necessary to transfer a determinate volume of fluid from the flask ; it is desirable, therefore, that the same flask should have a double graduation—one to *contain*, the other to *deliver*.

1,000 c.c. of distilled water at 4° C. weigh 1,000 grams. If, therefore, we place the litre-flask, perfectly clean and dry, on one pan of a balance capable of turning with 0·05 gram when carrying 2 kilos, and tare it, placing 1,000 grams on the weight-pan, and pouring in water of 4° C. until the equilibrium is established ; the level of the water will indicate to us the proper position of the graduating line. The flask *contains* 1,000 c.c. of liquid when filled up to that line. If we now pour out the water, allow the flask to drain for a few seconds, remove the 1,000 grams from the weight-pan, and re-adjust the tare of the flask, replace the 1,000 grams, and again fill up the flask with water at 4°, until the equilibrium is again established, the level of the water will now indicate to us the position of what we may call the *delivery-mark*. The flask filled up to this mark and emptied, *delivers* 1,000 c.c. But a very superficial observance of the surface of the liquid in the neck of the flask shows us that it is not perfectly horizontal. Unless, therefore, we invariably make some determinate point of the curve to coincide with the graduating line, our measurements will not be uniform. It will be found most convenient to take the lowest point of the curve or *meniscus* as the fixed point. In verifying or correcting the graduation of the flask, the true mark is scratched with a diamond so as to coincide with the lowest point of the curve of water in the neck ; and when it is desired that the flask shall be filled with 1,000 c.c., the liquid is to be poured in until the lowest portion of its surface exactly reaches this position.

The distilled water in a laboratory has very seldom a temperature of 4°, but as we know from experiment the rate at which the liquid expands, it is easy to calculate what would be the weight of 1,000 c.c. at any given temperature. This weight may be obtained from the following table.

The weight of 1,000 c.c. of water of $t°$ C., when determined by means of brass weights in air of 0° C., and of a tension 0·76m., is equal to $1,000 - x$ grams.*

$t°$	0	1	2	3	4	5	6	7	8	9
x	1·25	1·20	1·15	1·13	1·12	1·12	1·14	1·16	1·21	1·27

* Watts's 'Dictionary of Chemistry,' vol. i. p. 256.

$t°$	10	11	12	13	14	15	16	17	18	19
x	1·34	1·43	1·52	1·63	1·76	1·89	2·04	2·20	2·37	2·55

$t°$	20	21	22	23	24	25	26	27	28	29
x	2·74	2·95	3·17	3·39	3·63	3·88	4·13	4·39	4·67	4·94

The student is now in possession of all the data required to graduate his measuring vessels. He should fill a large beaker with distilled water, place it in the balance-room, and ascertain its temperature. Let us suppose that it is 15°, and that he requires to graduate his litre-flask to *contain* 1,000 c.c. On reference to the table, we see that the value of x corresponding to 15° is 1·89; accordingly, the weight of water necessary to be poured into the flask is 1,000 − 1·89 = 998·11 grams. In graduating the 250 c.c. flask, he would of course take one-fourth of this amount, viz., 249·53 grams. He places 998·1, or 249·5 grams, as the case may be, on the weight-pan, in addition to the tare of the flask, and fills up the flask with water until it is exactly equipoised. He then marks with a diamond on the neck of the flask the position of the lowest point of the meniscus. He now repeats the observation in the manner already described in order to obtain the graduation for *delivery*.

He next proceeds to re-graduate his pipettes. The light frame A B (fig. 37), made of stout brass wire, carries two clips of thin sheet brass closed by sliding collars ; through the lower clip is inserted the upper end of the pipette to be graduated : this is connected by caoutchouc tubing with the glass stopcock C, to which a short length of thermometer tube can be attached, as shown in the figure. To begin with, the thermometer tube is removed, and a piece of wider glass tubing placed in the caoutchouc tubing, and, the stopcock being opened, the pipette is filled by suction with distilled water a centimetre or so above the mark. The object of the glass tube is to prevent the caoutchouc being moistened by the lips. The end of the pipette which has been dipped beneath the surface of the water is dried by a cloth, the stopcock is reopened, and the water is allowed to flow out again

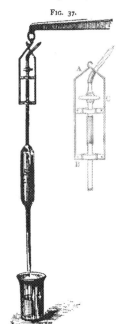

Fig. 37.

by its own weight into the beaker. As soon as the flow of water has ceased, the pipette is held vertically for three or four seconds to allow the liquid adhering to the glass to flow down into the stem; the end is then caused to touch the surface of the water. Of the various methods of delivering pipettes, this is most accurate. The glass tube is withdrawn from the caoutchouc, and the short length of thermometer tubing, the end of which is drawn out before the lamp so as to make the bore of very small diameter, is placed in its stead. The whole is then suspended from the arm of a balance turning with 0·05 gram, in the manner represented in fig. 37, and accurately counterpoised. The pipette is removed from the balance, the thermometer-tube is withdrawn, and the wide glass tube reinserted; the cock is opened, and the water is again drawn into the pipette one or two centimetres above the mark already etched upon the stem. The end of the pipette is again wiped with a dry cloth, the glass tube is replaced by the thermometer tubing, and the pipette is again suspended from the balance arm. The temperature of the water is observed; suppose it to be 15°, and that the pipette is to deliver 50 c.c., we find from the table that the weight of water possess-

ing this volume is $\dfrac{1000 - 1\cdot89}{20} = 49\cdot90$ grams. This weight
is accordingly placed on the weight-pan, in addition to the
tare, and the balance is caused to oscillate. In all probability the pipette and its contents will be too heavy ; the
cock is now opened, and one or two drops of water are
allowed to flow out into a beaker placed below. On
account of the slowness with which the air finds it way
through the narrow bore of the thermometer-tube, the
number of drops may be regulated with great nicety.
Successive drops are thus allowed to flow out until the
balance is in equilibrium. The lowest part of the meniscus
is then marked on the stem. The pipette will now deliver
50 c.c., if emptied in the manner described. The determination should be repeated ; if made with proper care, the
level obtained in the second experiment will be identical
with that found in the first. The capacities of the remaining pipettes are verified in the same manner.

The Burette.—This instrument serves to deliver definite
volumes of the standard solutions. Various forms of the
burette have been devised, but the most convenient modifications are those of Gay-Lussac and Mohr. Gay-Lussac's
burette is seen in fig. 36 *c.* It consists of a tube about
30 centimetres long, and 1·4 to 1·8 centimetres wide, sealed
at one end, and furnished with a narrow side-tube, starting
near the bottom, and running close to the side, to within
about 2 centimetres from the open end, where it is bent
slightly in the manner seen in the figure. These burettes
are usually made in two sizes—one to hold 25 c.c., and
graduated in $\frac{1}{10}$ c.c. ; the other to hold 50 c.c., and graduated in $\frac{1}{5}$ c.c. They are graduated 'for delivery.' The
correctness of the graduation should be tested previous to
use, by filling the burette with distilled water, and emptying
it through the side-tube, until the bottom of the meniscus
is coincident with the lowest division. The temperature of
the water is then ascertained, the burette is tared, and the

weight of the water supposed to be required to fill it is
placed on the weight-pan, and the instrument is filled up

FIG. 38.

FIG. 39.

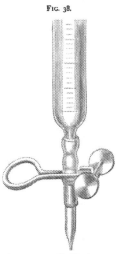

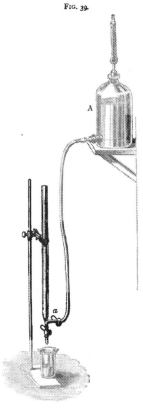

with the distilled water
until the balance is in equilibrium ; if the meniscus is
now coincident with the
zero point, the instrument
is correctly graduated. Differences of less than 0·05
c.c. may generally be neglected. So long as its bore
is uniform, and the divisions
are of an invariable width,
the instrument need not be
discarded, even if the lowest point of the meniscus is not
coincident with the zero. Let us suppose that on a 50 c.c.
burette, on which 250 divisions=50 c.c., the lowest point of

the meniscus was coincident with the division corresponding to 0·8 c.c., then obviously 250−4, or 246 divisions, are
equivalent to 50 c.c., and 1 division$=\frac{50}{246}$ c.c., or 0·203 c.c.
Accordingly, the indications of the burette must be multiplied by 1·016, to give the correct number of cubic centimetres
delivered. Thus, if in an analysis we had delivered apparently 25 c.c. from such a burette, we should in reality
have delivered 25 × 1·016 = 25·4 c.c. of liquid.

In using the burette, the edge of the side-tube should
be greased slightly ; this prevents the possibility of liquid
adhering to the outside of the tube when the burette
is replaced vertically in its support. With a little practice it
is easy to deliver the liquid in a stream or in drops ; when
the burette is brought to the vertical, to be read off, it is
necessary to wait for a
few seconds before
making the observa-
tion, in order that the
liquid may attain a con-
stant level.

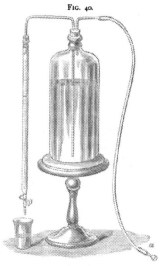

FIG. 40.

Mohr's burette is
seen in fig. 38. It is
simply a divided tube,
contracted at its lower
end, and fitted with a
short length of caout-
chouc tubing into which
is inserted a glass jet.
The sides of the caout-
chouc tube can be
pressed together by
means of the spring
clamp. This form of
burette is not so gene-
rally applicable as that
of Gay-Lussac, since the caoutchouc is acted upon by

several of the substances employed in standard solutions. In
the more modern form of the burette, a glass stop-cock is substituted for the india-rubber and clamp. This modification
(fig. 36*a*) leaves nothing to be desired. It is especially con·
venient where a great number of analyses of the same kind
have to be made, as in metallurgical laboratories, chemical
works, &c. In such cases the burette may be conveniently
arranged as shown in fig. 39. The bottle A contains the
standard solution; on opening the clamp *a*, the liquid fills
the burette gradually, and without the formation of air-
bubbles. Fig. 40 shows another method of connecting the
burette with the reservoir of the standard solution. The
liquid is driven into the burette by simply blowing through
the caoutchouc tube *a*. The graduation of Mohr's burette
may be verified by filling the instrument with water, and
allowing successive quantities of, say, 10 c.c., to flow out
into a weighed beaker. If the 10 c.c. weigh 9·98 grams, the
burette is correctly graduated. Of course due care must be
taken to allow the liquid adhering to the sides of the tube
to flow down before the level is read off.

The correct *reading off* of the burette may be facilitated
by the use of a little device recommended by Mohr. A
broad strip of black paper is pasted on a white card or
sheet of white paper, and this is held behind the burette,
so that the edge of the black paper is about 2 mm.
below the dark zone of the liquid. The lower edge of the
liquid is thus sharply defined, and may be read off with
certainty. A little caoutchouc band, slipped round the tube,
and through the card, renders the arrangement more convenient.

In reading off the Gay-Lussac burette, the level of the
liquid should be brought to the direct line of vision. This
may conveniently be determined by pasting a narrow strip
of black paper upon the side of the room, ten or twelve feet
from the operator, and on a level with his eye. The burette
is held perpendicularly between the thumb and first finger.

in such position that the black strip appears immediately behind the level of the liquid. Greater certainty in reading off may be attained by the use of Erdmann's float (fig. 41). It is simply an elongated glass bulb, somewhat smaller in diameter than the burette, containing a small quantity of mercury. The upper end is drawn out, sealed, and bent into a little hook, by which the bulb can be lifted in and out of the burette by the aid of a bent wire. Round the bulb runs a line *a*, etched by means of hydrofluoric acid, or scratched by a diamond. The coincidence of this line with the division of the burette is taken as the reading. The float should move easily within the burette, and so that the line is always parallel with the divisions; by its means the volume of the liquid delivered may be determined to within 0·005 c.c.

FIG. 41.

In certain cases the quantity of the liquid delivered is determined by weight. The solution is contained in the little weighed flask seen in fig. 42. The required amount is poured through the delivery-tube, which should be slightly greased at the edge. By weighing the apparatus before and after delivery, the amount of liquid employed is at once determined. A method of making a simple form of this apparatus is described under the section 'Ash Analysis.'

FIG. 42.

We now proceed to the experimental study of certain volumetric processes. We shall describe here a few typical processes to enable the student to familiarise himself with this mode of estimation. Other methods will be given in Part IV.

I. Determination of Chlorine by Standard Silver-Solutions.

If we add silver nitrate to a solution of a chloride, say of common salt, we obtain a white precipitate of silver chloride; if we continue to add the silver solution, the formation of this substance goes on until the whole of the chlorine is precipitated. If we add silver nitrate to a solution of potassium chromate, we obtain a dark-red precipitate of silver chromate. If now we mix the alkaline chloride and chromate together, and cautiously add, little by little, the silver nitrate solution, we notice that the chloride is first decomposed; and white silver chloride continues to be formed so long as any chlorine remains in solution. It is only after the whole of the chlorine is precipitated that we observe the formation of the dark-red silver chromate. This principle constitutes the basis of an accurate volumetric process. To carry it out we require a standard solution of pure silver nitrate free from excess of acid, and a solution of potassium chromate.

Preparation of Pure Silver.—Chemically-pure silver is frequently needed in volumetric analysis. We not only require it in the present process: we shall have occasion to use it in determining the strength of the hydrochloric acid solution employed in alkalimetry. It is therefore desirable that the student should prepare at one time all that he will need in this and subsequent operations.

About 50 grams of standard silver (composed of 12·3 parts of silver and 1 part of copper) are dissolved in dilute nitric acid in a thin porcelain basin at a gentle heat ; the solution is evaporated to dryness, and the residue heated to fusion. The cooled mass is then dissolved in ammoniacal water, allowed to stand for a short time, and filtered into a large flask. The filtrate is diluted to 2½ litres. 50 c.c. are withdrawn, heated nearly to boiling, and mixed with a solution of neutral ammonium sulphite (prepared by neutralising ammonia with sulphur dioxide gas) added drop by drop, until the liquid is decolourised. The ammonium sulphite solution is then mixed in the proportion demanded by this trial with the 2450 c.c. of liquid in the flask, which is then closed air-tight. In about 48 hours, nearly a third part of the silver is deposited as a crystalline powder, and the remainder is thrown down on heating the liquid to 60° or 70° for a short time. The liquid is now completely decolourised, unless it contains nickel or cobalt, which are not infrequent impurities in standard silver, when it will be light-green or pink. The precipitated silver is washed with distilled water, and digested with strong ammonia, again washed and dried. It is then fused with about 3 grams of ignited and powdered borax, previously mixed with a little sodium nitrate, in an unglazed porcelain crucible, and the button of metal is washed with hot water, and rubbed, if necessary, with a little sea-sand. It should then be rolled out into foil, sufficiently thin to be readily cut with a pair of scissors.

Preparation of the Standard Solution of Silver.—10·794 grams of the foil are weighed out, placed in a porcelain basin provided with a glass cover, and dissolved in dilute *pure* nitric acid on the water-bath. When the whole of the metal is dissolved, the under surface of the glass is rinsed into the dish, and its contents are evaporated to *complete* dryness on the water-bath, and gently heated over the lamp until the salt fuses. The dry and neutral silver nitrate is dissolved in pure water, and the solution carefully poured into the litre-flask, the dish being repeatedly washed out with fresh portions of distilled water. The flask is now filled up to the *containing-*mark with distilled water, the stopper is inserted, and the flask well agitated. The liquid constitutes a *deci-normal solution* of silver nitrate : 1 c.c. = 0·010794 gram silver ; it is therefore equivalent to 0·003546 gram of chlorine, or ·003646 gram of hydrochloric acid, or ·00585 gram of sodium chloride. The solution should be poured into a perfectly clean and dry bottle, provided with a well-fitting stopper : it should be labelled ' Deci-normal Silver Solution.' NOTE.— By a *normal solution* is to be understood a solution containing 1 eq. of the substance, in grams, dissolved in 1,000 c.c. of liquid. Thus a normal solution of silver would contain 107·94 grams of the metal in 1 litre of the solution. A normal solution of hydrochloric acid would contain 36·46 grams of HCl in 1,000 c.c. A deci-normal solution contains one-tenth of an equivalent; a centi-normal the one-hundredth part of an equivalent, in grams, per litre.

Preparation of Potassium Chromate Solution.—The commercial salt is recrystallised until it is free from chlorine : the solution acidified with nitric acid should not give the least turbidity on the addition of a drop of silver nitrate solution. Its solution should be kept in a little bottle A, through the cork of which runs a narrow tube with a mark *d* scratched upon it. This allows of a constant quantity of the solution to be withdrawn from the bottle. (Fig. 43.)

FIG. 43.

The Process.—A quantity of pure sodium chloride (see p. 79), is powdered, and gently heated, and whilst warm introduced into a small tube, fitted with a good cork. About 1 gram of the chloride is accurately weighed out into the ¼ litre flask, and dissolved in distilled water ; the flask is filled up to the *containing*-mark, and the solution well agitated. The burette (either Gay-Lussac's or Mohr's may be used) is rinsed out with a little of the standard silver solution (which is thrown into the ' silver residue ' bottle) and filled up to the zero with the silver solution. 50 c.c. of the solution of sodium chloride are withdrawn from the flask, and run into a porcelain basin, and mixed with a measure of the chromate solution ; the silver solution is added, drop by drop, from the burette, until the red colour of the silver chromate is permanent. Each drop of the silver solution forms a red spot in the yellow liquid, which quickly disappears, so long as any chloride remains in solution ; immediately all the chlorine is precipitated, the red colour of the chromate of silver is unaltered. The process is now at an end. The volume of the silver solution employed is read off, corrected, if necessary, for the error of the graduation (see p. 121), and 0·1 c.c. subtracted, this expenditure of silver being required to render the final reaction evident. The analysis should be repeated on a second portion of 50 c.c. of solution. An actual example will render the method of calculation clear. 1·0850 gram of pure salt was dissolved in 250 c.c. of distilled water. 50 c.c. of this solution required in experiment I., 37·1 c.c.; in experiment II., 37·2 c.c. ; in experiment III., 37·2 c.c. of silver solution. Mean 37·17 c.c. Subtract 0·1 for final reaction. 37·07 × ·003546 = 0·1314 gram of chlorine. 50 c.c. of liquid contain 0·217 gram of salt. Accordingly, the salt contains $\frac{0·1314 \times 100}{0·217} = 60·56$ per cent. chlorine. Theory requires 60 60 per cent.

The remainder of the solution of the sodium chloride should be poured into a clean and dry stoppered bottle, and its strength marked on a label attached to the bottle. It will be useful in cases where, in determining chlorine by this method, we imagine that we have added an excess of silver solution. We have only to add a definite volume, say 1 c.c. of the solution, to the turbid liquid, and after the last trace of silver chromate has disappeared, again add the silver solution until the final point is exactly obtained. In the case above cited, 50 c.c. of salt solution equal 37·07 c.c. of silver solution ; accordingly 1 c.c. of salt = 0·7 c.c. of silver. 0·7 + 0·1 (for final reaction) or 0·8 c.c. subtracted from the total amount of silver solution employed, gives the exact amount used in the analysis.

II. INDIRECT DETERMINATION OF POTASSIUM AND SODIUM BY MEANS OF STANDARD SILVER SOLUTION AND POTASSIUM CHROMATE.

The deci-normal silver solution may be used for a variety of estimations in which the amount of chlorine present may be taken as a measure of the other constituents. We shall have occasion to mention several of the applications of this solution in the General Part (Part IV.). We have already described the method of estimating potassium and sodium by gravimetric analysis: as an example of the above-mentioned applications of the solution of standard silver, we proceed to show how these alkalies, when together, may be estimated by volumetric analysis.

From 3 to 4 grams of pure Rochelle salt ($C_4H_4KNaO_6$. $4H_2O$) are gently heated in a platinum basin until the water of crystallisation is expelled. The temperature is then increased until the mass is completely carbonised; the heat should not exceed low redness, or a loss of alkali will be incurred. The alkaline carbonates in the charred mass are then dissolved in a small quantity of hot water, filtered, and the charcoal repeatedly washed with successive quantities of water. A slight excess of pure hydrochloric acid is added to the filtrate contained in a weighed platinum basin, and covered with a watch-glass, and as soon as the evolution of gas ceases, the under surface of the watch-glass is rinsed into the basin, and the liquid is evaporated to *complete* dryness, and heated in the air-bath to 180°. The alkaline chlorides are weighed and dissolved in a small quantity of water, the solution poured into a $\frac{1}{4}$-litre flask, and diluted to the *containing*-mark. 50 c.c. are then withdrawn and titrated with silver solution and potassium chromate in the manner described. From the weight of the chlorides, and of the chlorine they contain, we can readily calculate the proportion of the alkalies in the mixture.

Let x stand for the potassium, and y for the sodium, s for the weight of the mixed chlorides, and A for that of the chlorine found.

$$x = \frac{[(s \to A).\ 1\cdot54] - A}{0\cdot63}$$

$$y = \frac{A - [(s - A)\ 0\cdot91]}{0\cdot63}$$

$$1\cdot54 = \frac{Cl}{Na} \qquad 0\cdot91 = \frac{Cl}{K} \qquad 0\cdot63 = \frac{Cl}{Na} - \frac{Cl}{K}$$

III. ESTIMATION OF CHLORIC ACID.

Weigh out about 0·5 gram of dry potassium chlorate into a small beaker in which you have previously placed about 20 grams of thin sheet-zinc covered with spongy copper, in the manner described on p. 96. Add about 25 c.c. of water, cover the beaker with a watch-glass and boil the liquid gently for about an hour. Add water to the beaker, filter the liquid into a porcelain basin, and wash the zinc and copper in the beaker repeatedly with hot water. By the action of the nascent hydrogen, the alkaline chlorate is reduced to chloride. The filtrate should be quite neutral. Determine the amount of chlorine in the liquid by standard silver and potassium chromate solutions. 1 c.c. of the solution is equivalent to 0·01226 gram of potassium chlorate.

Example.—0·2492 gram of potassium chlorate, treated in the manner described, required 20·2 c.c. of deci-normal silver solution. 20·2 × ·01226 = 0·2477 gram potassium chlorate.

IIIA. DETERMINATION OF CHLORINE IN PRESENCE OF SULPHITES.

Add to the solution of the salts a very slight excess of a solution of potassium permanganate free from chlorine ; neutralise the liquid with pure soda, and then add the potassium chromate and standard silver nitrate solution in the usual manner. This preliminary oxidation of the sulphurous acid is necessary, otherwise the chromate solution would be reduced.

ALKALIMETRY.

Preparation of Normal Solution of Hydrochloric Acid.—1 c.c.
=0·03646 gram HCl.   Messrs. Roscoe and Dittmar have
shown that if a solution of hydrochloric acid containing 20·2
per cent. HCl be boiled under the ordinary pressure of the
atmosphere, the acid and water distil over in the proportion
in which they are contained in the boiling liquid.   If we take
a solution of the acid having approximately this composition
and boil it in a retort until about half of it has distilled over,
we may be sure that the residue contains about 20·2 per cent.
of acid.   This principle affords the basis of a method of
preparing a standard solution of hydrochloric acid.

We commence by ascertaining the specific gravity of a
strong solution of hydrochloric acid by means of the hydrometer (see Appendix), and we then add water to it until its
specific gravity is reduced to 1·1.   A solution of this strength
contains about 20·2 per cent. of HCl.   The amount of water
x which we require to add to a measured quantity of strong
hydrochloric acid A, of specific gravity a, to reduce it to the
specific gravity b (in this case 1·1), is found from the formula

$$x = \frac{A\,(a - b)}{b - 1}$$

Let us suppose that the specific gravity of our acid is 1·16:
in order to bring its specific gravity down to 1·1 we shall
require to add to every 100 c.c. of acid $\dfrac{100\,(1\cdot16 - 1\cdot1)}{1\cdot1 - 1} =$
60 c.c. of water.   500 c.c. of strong acid are mixed with the
quantity of water required to reduce its specific gravity to 1·1,
and the mixture is brought into a retort connected with a good
condensing arrangement, and boiled until nearly one-half the
amount has distilled over.   The ebullition may be rendered
more regular by throwing a few scraps of clean platinum foil
into the liquid.   The residue contains about 20·24 per cent.
of HCl.   180·8 grams of such acid, when diluted to a litre,
furnish a solution which is approximately normal.

Since this acid is of frequent application, the student should prepare from 2 to 3 litres of its solution. To determine its exact strength, 50 c.c. of the acid solution are run into the ½-litre flask, and diluted to the *containing*-mark after shaking. 1 c.c. approximately equals 0·003646 gram HCl. Call this solution A; 25 c.c. of A are further diluted to 250 c.c. 1 c.c. = ·0003646 gram HCl. Call this solution B.

Weigh out exactly 1·0794 gram of pure silver into a bottle of about 300 c.c. capacity, provided with a well-fitting stopper, and dissolve the metal in pure dilute nitric acid. The solution should be heated on the water-bath, and the fumes of the oxides of nitrogen should be blown out of the bottle from time to time. When the silver is completely dissolved, and the liquid on agitation gives no trace of red fumes, the bottle is removed from the bath and allowed to cool. A 100 c.c. pipette is rinsed with a small quantity of solution A, which is allowed to flow away. The pipette is filled to the mark with the solution A, and emptied into the silver solution. The stopper is inserted, and the solution is briskly agitated for some time until the silver chloride settles out completely and leaves the liquid almost clear. If the hydrochloric acid is of exact strength, that is, if it is strictly normal, and if 1·0794 gram of pure silver has been accurately weighed out, we ought of course to have neither silver nor chlorine in excess in solution. In all probability we shall have one or other of the bodies in excess. To determine which of the two remains in solution, we add 1 c.c. of deci-normal silver solution and note whether a further turbidity ensues. If the liquid remains clear, the silver is in all probability already in excess: if it becomes turbid it is a sign that the chlorine is present in excess. In the latter case the solution is again vigorously agitated until the liquid is once more clear. We will assume, by way of example, that the addition of the 1 c.c. of deci-normal silver solution produced a turbidity. Another 1 c.c. of silver solution is added to the liquid, and we again observe whether a turbidity is caused. Let us suppose that the liquid now remains clear : it is evident that an excess of silver is in solution. The total amount of silver in the bottle is therefore 1·0794 + (·010794 × 2) = 1·100988 gram. If we determine the amount in solution, we can at once tell how much is precipitated as chloride, and accordingly calculate the amount of HCl in the 100 c.c. of diluted acid. It is evident that we have at least 0·010794 gram of silver in excess, since the addition of the last 1 c.c. of deci-normal silver solution failed to produce a turbidity. We now add 1 c.c. of solution A, and shake the liquid vigorously until it is clear. In all, we have added 101 c.c. of solution A, and 1·100988 gram silver. We now add 1 c.c. of solution B, and note whether a turbidity is produced ; if so, we again shake the liquid vigorously until it is clear, and add a second 1 c.c. : if the liquid is still rendered turbid, we again shake briskly, and add a third 1 c.c., and so on until the addition of 1 c.c. of solution B no longer produces any change. This last c.c. is not counted, since it shows that the HCl is once more in slight excess, and we shall be nearer the truth if we assume that only half of the preceding c.c. is necessary for precipitation. In working with the centi-normal solution (solution B) it is necessary that the liquid above the precipitated silver chloride be perfectly clear, and that we wait for a few seconds (say $\frac{1}{2}$ minute) before we conclude that no further turbidity is caused by the addition of 1 c.c. of solution. Let us suppose that we found it necessary to add 5 c.c. of solution B before the liquid remained clear : 3·5 c.c. are therefore necessary to precipitate the silver in solution after the addition of the 1 c.c. of solution A.   If the directions given have been properly followed, we may assume without sensible error that 1 c.c. of solution B contains 0·000365 gram HCl : this is equal to ·0010794 gram Ag., and ·0010794 × 3·5 = 0·003778 gram Ag.   101 c.c. of solution A are equivalent to 1·100988 − ·003778 = 1·09721 gram Ag., and accordingly

$$107·94 : 36·45 :: 1·09721 : x$$
$$x = 0·370514.$$

101 c.c. of solution A contain 0·370514 gram HCl. Accordingly the 500 c.c. would contain 1·83423 gram HCl. But the 500 c.c. of A are equivalent to 50 c.c. of the original acid. Accordingly 1 c.c. of the original acid contains ·036684 gram HCl, instead of ·03645 gram, the quantity required to constitute the normal acid. Instead of diluting the acid to bring it to the exact strength, it is better to express the difference by a small factor: in this case $\dfrac{·036684}{·03645} =$ 1·0064. The acid is accordingly labelled ‘ *Standard Hydrochloric Acid.* 1 *c.c.* = 0·03645 × 1·0064 *gram HCl.*’ 1 c.c. of the acid is equivalent to 0·10794 × 1·0064 gram Ag., or 0·04004 × 1·0064 gram NaHO.

At least two determinations should be made of the strength of the acid before it is used, and the mean result should be taken as indicating the correct value.

Preparation of Normal Sulphuric Acid Solution.—In certain processes the use of standard hydrochloric acid is inadmissible ; in such cases we may generally employ a normal solution of sulphuric acid. This may be prepared by diluting about 60 c.c. of concentrated and pure sulphuric acid with five or six times its volume of water, allowing the mixture to cool, and making it up to 2 litres. If the sulphuric acid used was concentrated, the solution will now contain rather more than 49 grams H_2SO_4 per litre. To determine its exact strength, weigh out about 2 grams of recently heated pure sodium carbonate into a weighed platinum basin, dissolve it in a small quantity of water, cover the solution with a watch-glass, draw the watch-glass aside and add 25 c.c. of the acid. Place the liquid on a water-bath, and as soon as the evolution of gas has ceased, remove the cover, rinse its under surface into the dish, and evaporate the liquid to complete dryness. Heat to 180° in the air-bath until the weight is constant. The calculation is very simple. $SO_4 = 96$ has displaced $CO_3 = 60$. The increase in weight of the dish is proportional to the amount of sulphuric acid employed—so long of course as there is excess of sodium carbonate present. The amount of sulphuric acid x in the 25 c.c. is thus found :—

The difference between the equivalent of SO_4 and CO_3, viz. 36, is to the equivalent of SO_4H_2, viz. 98, as the difference between the first and second weighing of the platinum basin is to the sulphuric acid present in the 25 c.c.

Example.—Weighed out 1·7210 gram of pure dry sodium carbonate, added 25 c.c. of the acid, evaporated to complete dryness, and heated in the air-bath. The difference between the weight of the dish + carbonate, and dish + mixed sulphate and carbonate was 0·465 gram : then

$$36 : 98 :: 0·465 : x$$
$$x = 1·266 \text{ and } \frac{1·266}{25} = 0·05064.$$

1 c.c. of the acid accordingly contained 0·05064 gram H_2SO_4.

This method of determining the exact strength of the acid solution is quite as accurate, and certainly more convenient, than precipitating the sulphuric acid as barium sulphate. By way of control the acid solution was treated with barium chloride and the precipitate washed, dried, and weighed. 10 c.c. of the solution gave 1·2043 gram $BaSO_4$ as the mean of three concordant experiments. This is equal to 0·5063 gram of sulphuric acid. The determination by sodium carbonate gave 0·5064 gram. The determination of the strength of the acids employed in volumetric analysis may, in many cases, be accurately and expeditiously made by means of metallic sodium. About 0·5 gram. of clean freshly-cut sodium is placed in a short wide-mouthed test-tube, previously weighed, together with its well-fitting cork ; the tube containing the sodium is corked and again weighed ; the metal is thrown upon about 15 c.c. of cold water contained in a porcelain basin, which is then quickly covered with a large watch-glass. When the action is at an end, the glass is raised, a few drops of litmus solution (*vide infra*) added, and the acid is run in from a burette until the red tint is permanent.

The determination of the strength should be repeated once or twice, and the mean result taken as expressing the true value of the acid. If the solution is approximately normal it is better not to dilute it, but to calculate the factor required to bring its strength to the normal value. Thus the factor of the above-mentioned solution would be $\frac{50.64}{49} = 1.0334$. It would be labelled therefore '*Normal Sulphuric Acid.* 1 *c.c.*='049 *gram* × 1·0334 SO_4H_2.'

If the acid is much above the normal value it will be more convenient to dilute it so as to make it as nearly as possible of the proper strength. Thus, supposing that we had found that 1 c.c. contained 0·055 gram H_2SO_4, 1000 c.c. would contain 55 grams. Consequently, according to the proportion

$$49 \,:\, 1000 \,::\, 55 \,:\, x. \quad x = 1123.$$

we must add 123 c.c. of water to every 1000 c.c. of the acid solution. This may be best effected by filling the litre flask up to the *containing*-mark with the acid solution, and emptying it into the dry and clean bottle in which it is to be preserved. Now pour into the litre flask 123 c.c. of water (by means of the 100 c.c. pipette and the burette), shake the liquid about in the flask and pour it into the bottle, and shake the mixture. Again pour about half the acid liquid back into the litre flask, shake and transfer it once more to the bottle.

Preparation of Normal Caustic Soda Solution.—Dissolve from 42 to 45 grams of sodium hydrate in 800 c.c. of water, and titrate the solution by normal acid and litmus tincture. The alkaline solution is then diluted until it possesses the normal strength.

The caustic soda solution may also be obtained by dissolving about 150 grams of pure dry sodium carbonate in 3 litres of water, boiling the solution in a clean iron vessel, and adding, little by little, 80 grams of freshly-burnt lime made

into a cream with water.* The mixture must be boiled until
a small quantity of the clear solution no longer effervesces
on the addition of an acid in excess. The iron vessel is then
closely covered, and after standing 12 or 14 hours the clear
alkaline solution is drawn off by the aid of a syphon.

Preparation of Litmus Solution.—5 or 6 grams of coarsely-
powdered litmus are digested with about 200 c.c. of distilled
water for a few hours. The clear solution is decanted from
the sediment, and very dilute nitric acid added, drop by drop,
until the colour is changed to violet. The solution must be
neither red nor blue, but between the two in colour ; when
properly neutralised less than $\frac{1}{10}$ c.c. of the standard acid
should distinctly redden the solution of 1 c.c. in 100 c.c. of
water ; on the other hand the same amount of standard alkali
should render the colour decidedly blue. The solution should
be kept in a wide-mouthed bottle, the cork of which is so cut
that the air has ready access
to the interior of the bottle,
otherwise the liquid quickly
loses its colour. Through the
cork is fitted a short tube on
which is a mark ; this tube
serves to deliver a determinate
volume of the litmus solution.

FIG. 44.

*Determination of the Strength
of the Caustic Soda Solution.*—
25 c.c. of the standard sulphuric acid solution are poured
into a porcelain basin, mixed
with a measure of litmus solution, and the alkaline liquid is
added drop by drop from a
burette until the colour is just
turned to blue. Repeat the

* If this quantity of pure dry sodium carbonate is not at hand, 250
grams of the bicarbonate are heated to dull redness in a platinum basin,
in small portions at a time, for ten or fifteen minutes, to expel the
carbon dioxide. The salt is then treated as above.

determination, take the mean of the two observations, and dilute the alkaline solution until it corresponds volume for volume with the standard acid. Thus, supposing you have found that 25 c.c. of acid required 22 c.c. of soda for neutralisation, you will require to add 3 c.c. of water to every 22 c.c. of lye, or each litre of alkaline liquid will require the addition of 136 c.c. of water.

The diluted liquid should be poured into a large bottle fitted with a syphon and wide tube as shown in Fig. 44. The wide tube is filled with soda-lime in small pieces to prevent the entrance of carbon dioxide. A thin layer of refined petroleum or paraffin oil poured on the surface of the liquid greatly tends to the preservation of its strength. The exact strength of the diluted liquid should then be determined by neutralising varying qualities, say 25 c.c., 30 c.c., and 50 c.c., of standard acid in the manner above described. The mean result of the observations should be taken as the true value.

IV. VALUATION OF SODA-ASH.

Soda-ash is a crude sodium carbonate; its value depends on the amount of available sodium carbonate which it contains. Its impurities, in addition to moisture, mainly consist of sodium hydrate, sulphate, chloride, silicate, and aluminate. It also not unfrequently contains sodium sulphide, sulphite, and thiosulphate.

Weigh out about 10 grams of the powdered sample into a weighed platinum crucible, and heat gently for twenty or thirty minutes over the lamp; place the crucible in the dessicator, and weigh when cold. The loss of weight gives the amount of *moisture* contained in the sample. Transfer the weighed salt to a beaker, wash out the crucible, and dissolve the salt in a small quantity of water, filter (if necessary) into the ½-litre flask, wash the filter thoroughly, and dilute to the containing-mark, and shake. Take out 50 c.c. of the solution, corresponding to 1 gram of soda-ash, pour the liquid into a flask, and add 25 c.c. of standard sulphuric acid, and boil the solution for some time until the

carbonic acid is expelled.   Add a measure of litmus solution, and standard soda-solution from a burette, drop by
drop, until the blue colour of the litmus is restored.

Example.—10·025 grams of soda-ash were heated, dissolved in water, and the clear solution made up to 500 c.c.
50 c.c. (corresponding to 1·0025 gram soda-ash) were transferred to a flask, 25 c.c. of standard sulphuric acid
(1 c.c.=·049 gram $SO_4H_2 \times 1·0204$) were added, the liquid
was boiled, mixed with a measure of litmus solution, and
standard soda added, drop by drop, until the blue colour
was restored.   10 c.c. of soda solution=9·8 c.c. of standard
sulphuric acid.   9·2 c.c. of the alkaline liquid were needed.

$$10 : 9·2 :: 9·8 : 9·0$$

Accordingly 25−9·0=16·0 c.c. of standard acid have been
used to decompose the 1·0025 gram of soda-ash.   But
1 c.c. of acid contains 0·049 × 1·0204 gram of sulphuric
acid ; this corresponds to 0·053 × 1·0204 of sodium carbonate.   The amount of sodium carbonate in the 1·0025
gram of original soda-ash is therefore ·053 × 1·0204 × 16·0
=0·865 gram, or in 100 parts 1·0025 : 100 :: 0·865
=86·3 per cent.

The value of pearl-ash may be determined in exactly the
same manner.   It must not be forgotten that dried potassium carbonate is very hygroscopic ; due expedition must,
therefore, be employed in weighing this body.   The crucible
should be closely covered during the operation.   The
½ equivalent of potassium carbonate is 69·1 ; accordingly,
the factor ·0691 is used instead of ·053 in the calculation.

V. Estimation of Alkaline Hydrate in presence of Carbonate.

The crude carbonates of soda and potash not unfrequently
contain notable quantities of hydroxide.   The alkaline lyes
used by paper, soap, and starch manufacturers also consist
of mixtures of carbonated and caustic alkalies.

To estimate the proportion of the two constituents, a definite quantity of the salt or solution, say 15 grams, or 50 c.c., is dissolved in water, and diluted to 250 c.c. The total amount of alkali is then determined, say in 50 c.c., in the manner described on p. 137, viz., by standard acid, litmus, and soda solutions. Take out 100 c.c. of the alkaline solution, pour it into a 300 c.c. flask, dilute with a little water, heat, and add solution of barium chloride so long as a precipitate is formed. The reactions which occur in the case of sodium carbonate and hydrate are:

$$2NaHO + BaCl_2 = 2NaCl + BaH_2O_2,$$
$$\text{and} \quad Na_2CO_3 + BaCl_2 = 2NaCl + BaCO_3.$$

Fill up the flask to the containing-mark, shake, and allow the precipitate to settle. Withdraw 100 c.c. of the clear liquid, pour it into a porcelain basin, add an excess, but in measured quantity, of standard hydrochloric acid, a measure of litmus solution, and then determine the excess of acid added by means of the soda-solution. Multiply the amount of hydroxide found by 7·5 ; this gives the amount of caustic alkali in the weight of the substance originally taken.

VI. ESTIMATION OF SODIUM CARBONATE IN PRESENCE OF POTASSIUM CARBONATE.

Sodium carbonate, on account of its cheapness, is sometimes employed to adulterate pearl-ash. The quantity of the admixed sodium salt may be estimated in the following manner. About 5 grams of the mixture are gently heated in a weighed platinum crucible for fifteen or twenty minutes; the loss on weighing gives the amount of *moisture* present. The dried mass is dissolved in a small quantity of water, and filtered, if necessary ; acetic acid is added in slight excess, the liquid is heated to expel carbonic acid, and mixed with a dilute solution of lead acetate so long as a precipitate of lead sulphate is formed. The liquid is filtered, and the excess of lead removed by a stream of sulphuretted hydrogen; the solution is again filtered into a 250 c.c. flask, and diluted to the containing-mark. 50 c.c.

of the liquid are then evaporated to dryness, with about 10 c.c. of dilute hydrochloric acid (1·1 sp. gr.) in a weighed platinum dish. The residual chlorides are dried and weighed; the relative proportion of the potash and soda is then determined by means of standard silver and potassium chromate, in the manner described on p. 128.

VII. Determination of Ammonia.

The quantity of free ammonia in solution may be determined, as in the case of caustic soda or potash, by means of standard acid and litmus solutions. A definite quantity of the solution, say 10 c.c., is transferred to a small tared flask. and weighed: its absolute weight and specific gravity are thus determined in a single operation. If the 10 c.c. weighed 9·0 grams, its specific gravity would of course be 0·9000, water being 1. The weighed quantity of the ammonia is then diluted with 6 or 8 times its bulk of water and titrated directly in the ordinary manner by the standard acid.

The operation of taking the specific gravity and weighing out the ammonia solution may be rendered more accurate,

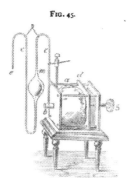

Fig. 45.

especially if the solution is very strong, by the aid of the little apparatus seen in Fig. 45. In the hole of the caoutchouc ball *a* is inserted a brass tube running through and fixed into the plate *b*. The end of the tube is closed, its upper edge is filed through, and over it is slipped a piece of tightly-fitting caoutchouc tube, in which, immediately over the orifice in the brass tube, is a hole pierced by a pin. Through the hole is inserted the end of the apparatus *cc*, which is further supported by the holder. The caoutchouc ball may be compressed by the plate *d*, moveable along the rods, by the aid of the milled-head screw s. On compressing the ball, the air makes its escape through the end *e* of the apparatus, and if, after the expulsion of the air, the end *e* be placed beneath the surface of a liquid, the liquid will be driven into the apparatus in proportion as the screw is reversed. If the apparatus has the arrangement seen in the figure, it is evident that it may be withdrawn from the caoutchouc tube without any of the liquid flowing out. If the capacity of the bulb up to a certain mark, say at *m*, be accurately estimated, by determining the weight of water it contains up to that mark, we can readily determine the specific gravity of any liquid introduced into it by simply weighing the apparatus filled with the liquid. By reinserting the end into the hole in the caoutchouc tube and compressing the ball we can deliver any required quantity of the liquid. On again weighing the apparatus, its loss of weight immediately gives us the amount of liquid delivered. The apparatus may also be used for weighing out and determining the specific gravities of strong acids, fuming liquids, &c.

Ammonia in combination may be determined by expelling it by means of caustic soda or lime, collecting the evolved ammonia in an excess of standard acid, and determining the excess of acid by soda-solution. The ammoniacal compound (say ammonium chloride) is weighed out into the retort *a*, Fig. 32, and the ammonia collected in excess of standard acid ; the amount of the residual acid is then determined by soda-solution.

The ammonia contained in many commercial salts, in ammoniacal gas-liquor, &c., may be determined by this method. In estimating the ammonia in guano, magnesia must be employed instead of lime or soda, otherwise the nitrogenous organic matter present will be partially decomposed with evolution of ammonia.

ACIDIMETRY.

The principles involved in the estimation of the strength of acid solutions are identical with those we have indicated under Alkalimetry. The value of the strong acids, such as hydrochloric, nitric, and sulphuric acids, is frequently deduced from their specific gravities, and comprehensive tables have been calculated showing the percentage amount of the various acids in solutions of different densities (see Appendix). Occasionally it is necessary to control the indications of the hydrometer, or specific-gravity bottle, by titrating the acid solution. The apparatus described on p. 140 (Fig. 45) may be conveniently employed to determine both the specific gravity of the liquid and the weight taken for analysis. The determination of the strength of nitric, hydrochloric, and sulphuric acids by caustic soda and litmus solution presents no difficulties : the method will be evident from the foregoing descriptions under Alkalimetry.

VIII.   DETERMINATION OF THE STRENGTH OF ACETIC,
ACID, PYROLIGNEOUS ACID, VINEGAR.

The estimation of the strength of this acid in its various forms cannot be made with very great accuracy by direct titration with caustic soda solution, since sodium acetate possesses a feeble alkaline reaction, which interferes with the correct determination of the final point. The method most generally applicable consists in adding to a known quantity of the acid, a weighed quantity (in excess) of finely-powdered marble, heating the liquid to boiling, filtering, washing the residual calcium carbonate with hot water, dissolving it in a slight excess of normal hydrochloric acid, and titrating with caustic soda and litmus solution. This method is particularly useful in testing brown pyroligneous acid or highly coloured vinegars.

[*Note.*—Instead of litmus solution, a dilute tincture of cochineal may be employed. It may be prepared by digesting 2 or 3 grams of powdered cochineal in 200 c.c. of a mixture of 1 part of alcohol and 4 of water. It forms an orange solution which is turned violet by alkalies : the colour is almost unaffected by carbonic acid.]

IX. DETERMINATION OF COMBINED CARBON DIOXIDE.

The amount of carbon dioxide in soluble carbonates may be readily determined by decomposing them with a solution of calcium chloride, throwing the precipitated calcium carbonate on to a filter, washing thoroughly with hot water, dissolving in an excess of standard hydrochloric acid, and determining the excess of acid by standard soda solution in the usual manner.

The acid carbonates (bicarbonates) require the addition of ammonia, with the calcium chloride.

The carbon dioxide in insoluble carbonates, as in calamine, ferrous carbonate, white lead, mortar, cements, &c., is determined by expelling the gas by the action of hydrochloric acid, absorbing it by ammonia, and precipitating by the addition of calcium chloride. The calcium carbonate is further treated in the manner above described.

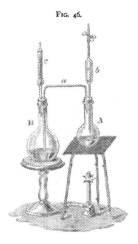

FIG. 46.

The decomposition may conveniently be effected in the apparatus seen in fig. 46. The flask A, of about 150 c.c. capacity, contains the weighed quantity of carbonate, together with about 10 c.c. of water : it is fitted with a caoutchouc cork, in which are inserted the bent tube *a* and the pipette-shaped tube *b*, filled with moderately-concentrated hydrochloric acid. The flow of the acid into the flask may be easily regulated by the clip. The flask B contains 10 or 15 c. c. of ammonia-water: the tube must not dip into the liquid. The wider tube *c* is partially filled with broken glass moistened with ammonia-water. Care should be taken that the ammonia is free from carbonic acid: its purity may be tested by adding to it a few drops of calcium chloride solution: if free from carbonate it will remain perfectly clear.

To make a determination, warm the flask B, so as to fill it with an atmosphere of ammonia, and then cautiously allow a few drops of hydrochloric acid to fall on to the weighed quantity of carbonate in A. As soon as the whole of the carbonate is decomposed, heat the liquid in A to boiling, so as to expel the last trace of carbon dioxide, and keep it boiling for a few minutes. Wash the bent tube *a* and also the glass in *c*, add calcium chloride solution to the ammoniacal liquid, boil for some time, filter, wash thoroughly, and titrate the calcium carbonate in the manner already described. The operation of filtering and washing should be done as expeditiously as possible, since the ammoniacal liquid absorbs atmospheric carbon dioxide.

X. Estimation of Carbonic Acid in Natural Waters.

The amount of carbonic acid in spring, river, or mineral water may be accurately estimated in the following manner. 100 c.c. of the water to be examined are transferred to a dry flask, together with 3 c.c. of a strong neutral solution of calcium chloride, and 2 c.c. of a saturated solution of ammonium chloride. 50 c.c. of lime-water, the strength of which is accurately known, are then added, the flask is closed by a caoutchouc cork, and its contents, amounting to 155 c.c., agitated. In about twelve hours the whole of the calcium carbonate will have separated out, and the liquid will be perfectly clear. 50 c.c. of the clear liquid are withdrawn, and the residual amount of lime determined by deci-normal hydrochloric acid.

A solution of oxalic acid may be conveniently substituted for the hydrochloric acid ; it should be made of such a strength that 1 c.c. is equivalent to 1 milligram of carbon dioxide. This solution may be obtained by dissolving 2·8636 grams of pure dry crystallised oxalic acid in water, and diluting to 1 litre. The lime-water should be of such strength that 25 c.c. are equal to 23 or 24 c.c. of acid. The final point of the reaction may be determined by the aid of a drop of tincture of pure rosolic acid, which gives a splendid red colour in presence of the alkaline earth, which disappears on neutralising with an acid. Instead of rosolic acid, turmeric paper may be used. Swedish paper, in strips, is immersed in tincture of turmeric, and dried. A drop of the liquid brought upon the paper gives a reddish-brown stain, so long as the least trace of the alkaline earth remains in the free state.

Example.—100 c.c. of the water of the Irish Channel were mixed with 3 c.c. of calcium chloride, and 2 c.c. of ammonium chloride, together with 50 c.c. of lime-water.

50 c.c. of lime-water$=46\cdot4$ c.c. of oxalic acid ;
of which 1 c.c.$=1$ milligram CO_2.

After standing fifteen hours—

50 c.c. of solution required 13·3 c.c. of standard acid for neutralisation.

Therefore—

$$46\cdot4-(13\cdot3 \times 3\cdot1) = 5\cdot2.$$

100 c.c. of the sea-water contain 5·2 milligrams carbon dioxide.

XI. Estimation of Carbon Dioxide in Artificially Aerated Waters.

The determination of the total quantity of carbon dioxide in artificial mineral waters,—seltzer,—soda-water, &c.—may be readily effected in the following manner. A narrow brass cork-borer is pierced with two or three small holes, about 4 or 5 centimetres from the edge. A piece of caoutchouc tubing is slipped over the upper end of the borer, and the other end is connected with the bent tube *a* of the flask B, fig. 47. The flask contains about 25 or 30 c.c. of moderately strong ammonia ; the end of the glass tube should dip beneath the surface of the liquid. The sides and edge of the brass tube are rubbed with a little paraffin, and it is then screwed through the cork of the bottle of the aerated water

Fig. 47

by holding the tube stationary, and turning the bottle round, until the holes make their appearance below the cork. The gas is immediately liberated ; it makes its escape through the holes in the borer, and is absorbed by the ammonia-water. The greater portion of the residual gas is expelled by shaking the water, and gently heating it by surrounding it with warm water. As soon as no more gas is evolved, the cork of the soda-water bottle is withdrawn, and the liquid added to that contained in the flask, the tubes are washed, calcium chloride is added, the liquid is boiled for some time, and the amount of calcium carbonate determined in the ordinary manner. The quantity of combined carbon dioxide in the water may be estimated by evaporating a second portion to dryness, gently igniting the residue, and titrating with standard acid and soda.

XII. Determination of Combined Acids in Salts.

If we add caustic soda to a boiling solution of copper sulphate, cupric oxide is precipitated, and sodium sulphate remains in solution :—

$$CuSO_4 + 2NaHO = CuO + Na_2SO_4 + H_2O.$$

If, therefore, we mix a measured quantity (in excess) of standard soda solution with a solution of copper sulphate, and boil, the amount of residual alkali indicates the quantity of the acid contained in the salt.

Weigh out about 2 grams of copper sulphate into a 250 c.c. flask, dissolve in 100 c.c. of water, boil, and add excess of standard soda solution, allow to cool, and dilute with water to the containing-mark, cork the flask, and allow the precipitate to settle. Take out an aliquot portion of the clear supernatant liquid, and determine the amount of residual alkali in solution. The method of calculating the result needs no explanation.

Many other substances, precipitable by sodium hydroxide or sodium carbonate, admit of estimation by this method. Thus we may determine the amount of acid present in

Silver salts, with caustic soda.

Mercury salts, with caustic soda.

Bismuth salts, with sodium carbonate.

Lead, nickel, cobalt, zinc, aluminium, manganese, alkaline earths, and magnesia, with sodium carbonate.

In certain cases the acid, as in copper sulphate, or mercuric chloride, may be liberated by treating a boiling solution of the salt with sulphuretted hydrogen, filtering, and determining the amount of the free acid by caustic soda and litmus solutions.

ANALYSIS BY OXIDATION AND REDUCTION.

We have already explained the main principle of this special form of volumetric analysis ; we have shown, for example, how the amount of iron in solution may be estimated by determining the quantity of oxygen required to convert it from the state of ferrous to that of ferric oxide. A large number of other substances may be estimated by the aid of a solution of a substance which, like potassium permanganate, readily parts with a portion of its oxygen. In all these cases the amount of oxygen given up is taken as an index of the quantity of the substance to be determined.

Of the many oxidising agents which are known, the most generally applicable are (1) potassium permanganate (permanganic acid), and (2) iodine.

A. ESTIMATIONS BY MEANS OF POTASSIUM PERMANGANATE.

Preparation of Potassium Permanganate Solution.—About 5 grams of pure crystallised potassium permanganate are dissolved in a small quantity of water, and the solution is diluted to 1 litre. It must be contained in a glass-stoppered bottle, which should be kept in a cool, dark place, when not in use. The solution thus preserved may be kept for a long time without experiencing much alteration.

XIII. DETERMINATION OF THE STRENGTH OF THE PERMANGANATE SOLUTION.

It is absolutely necessary to determine the power of the permanganate solution by direct experiment before using it; we cannot calculate its strength, i.e. the amount of oxygen that it is capable of furnishing, from the weight of the salt dissolved, on account of its instability. The most accurate method of estimating the strength of the solution consists in determining the amount required to transform a known weight of iron from the condition of ferrous oxide to that of ferric oxide.

About 1 gram of fine iron-wire (piano wire), perfectly free from rust, is accurately weighed out into a ½-litre flask, and dissolved in about 100 c.c. of dilute sulphuric acid (1 pt. of acid to 6 of water). A small quantity of sodium carbonate is thrown into the liquid at the same time in order that the air within the flask may be displaced by carbon dioxide. The flask is fitted with a cork and bent tube furnished with

a clip ; the end of the tube dips beneath the surface of about
25 c.c. of water contained in a small flask (fig. 48).
Whilst the iron is dissolving, the clip is kept open by
slipping it over the glass tube. The solution of the iron
may be accelerated by a gentle heat ; the liquid is gradually
caused to boil, and maintained in brisk ebullition for a
minute or two, so as to expel the mixture of carbon dioxide
and hydrogen, and the caoutchouc tube is immediately
closed and the lamp removed. In a minute or so the clip
is again opened, when the water from the little flask is driven
over into the solution of iron : in proportion as it passes
over, boiling water is poured into the small flask until the

FIG. 48

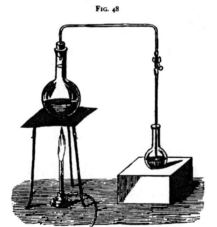

larger one is nearly filled. The caoutchouc tube is once
more closed, the flask and its contents allowed to cool
perfectly, and the volume of the liquid made up to the containing-mark, the stopper of the flask is inserted, and the
liquid thoroughly mixed by shaking. Whilst the solution is
cooling, fill a Gay-Lussac's or a Mohr's burette fitted

with a glass stop-cock (the permanganate solution gradually attacks caoutchouc), previously rinsed out with a little of the permanganate solution, read off the level of the permanganate solution, take out 50 c.c. of the iron solution, and pour it into about 200 c.c. of water contained in a beaker, standing on a sheet of white paper. Add the permanganate drop by drop to the liquid, with constant stirring, until the pink colour of the solution is permanent. The permanganate is at first decomposed with great rapidity : as the iron becomes oxidised the colour disappears more slowly; the rapidity of the change indicates the progress of the oxidation. The operation of standardising should be repeated once or twice on successive portions of 50 c.c. of iron solution, and the mean of the observations taken as representing the true value of the permanganate solution.

$$10FeSO_4 + 2KMnO_4 + 8SO_4H_2$$
$$= 5Fe_2(SO_4)_3 + 2MnSO_4 + K_2SO_4 + 8H_2O.$$

Let us suppose that we have weighed out 1·1 gram of wire and dissolved it in 250 c.c. of liquid : 50 c.c. of the solution would be equivalent to 0·2200 gram of iron. The iron we have taken, however, is not chemically pure : we may assume without sensible error that its impurities amount to 0·4 per cent. ; accordingly the amount of iron in the 50 c.c. is

$$1 : 0·996 :: 0·2200 : x. \qquad x = 0·2192.$$

Let us further suppose that we have required 20 c.c. of permanganate solution, as the mean of the experiments, before we obtained a permanent coloration with the 50 c.c. of iron solution : then 20 c.c. permanganate oxidise 0·2192 gram iron from protoxide to peroxide, or 100 c.c. permanganate are equivalent to 1·096 gram iron.

The strength of the permanganate solution may also be determined by means of pure ferrous sulphate, precipitated from its aqueous solution by means of alcohol. The ferrous sulphate so prepared keeps unchanged for years. Or, instead of this salt, the double sulphate of iron and ammonium $FeSO_4(NH_4)_2SO_4 + 6H_2O$ may be employed. It contains exactly one-seventh of its weight of iron: 0·7 gram of salt is equivalent to 0·1 gram of iron.

The strength of the solution may also be estimated by means of oxalic acid. If potassium permanganate solution is dropped into a solution of oxalic acid acidulated with sulphuric acid, the oxalic acid is completely decomposed into carbon dioxide and water:

$$5C_2H_2O_4 + 2KMnO_4 + 3H_2SO_4$$
$$= K_2SO_4 + 2MnSO_4 + 10CO_2 + 8H_2O.$$

The oxalic acid solution requires to be gently heated (to about 60°) before the reaction commences: at this temperature the permanganate is rapidly decomposed so long as any oxalic acid remains in the solution. From the foregoing equations it is evident that 112 parts of iron (in the state of protoxide) and 126 parts of crystallised oxalic acid $(C_2H_2O_4 + 2H_2O)$ require exactly the same amount of oxygen, viz. 16 parts, for complete oxidation. The amount of available oxygen in 1 c.c. of the permanganate solution may therefore be readily calculated.

Of the several substances which may be used for titrating the permanganate solution, metallic iron is on the whole to be preferred. All the methods are equally accurate with careful manipulation, but fewer sources of error attend the use of metallic iron. Crystallised oxalic acid is not readily obtained quite pure, it is liable to part with a portion of its water of crystallisation, and its solution, especially under the influence of light, is apt to decompose. Ammonium oxalate is preferable to oxalic acid : it may be readily purified by recrystallisation, and keeps perfectly well : its composition is $C_2O_4(NH_4)_2 + H_2O$. 142·08 parts of the crystallised salt are equivalent to 112 parts of iron.

Whichever method be adopted it is absolutely necessary that the solution to be titrated should contain free sulphuric acid. If there is a deficiency of free acid, the solution becomes brown, and eventually a precipitate is formed. It is not a matter of indifference which acid is employed for acidulation : nitric acid cannot well be used under any circumstances, and hydrochloric acid is liable to be decomposed, and chlorine eliminated, in accordance with the equation :

$$14HCl + Mn_2O_7 = 2MnCl_2 + 5Cl_2 + 7H_2O.$$

Whenever, therefore, the use of sulphuric acid is inadmissible, it is better to employ potassium bichromate as an oxidising agent (see Analysis of Iron Ores).

XIV. VOLUMETRIC ESTIMATION OF CALCIUM BY MEANS OF POTASSIUM PERMANGANATE.

Oxalic acid or ammonium oxalate solution added to calcium chloride or any soluble salt of lime, in presence of ammonium chloride and free ammonia, gives rise to a precipitate of calcium oxalate. Calcium oxalate digested with dilute sulphuric acid is decomposed, calcium sulphate is formed, and oxalic acid passes into solution. We already know that it is possible to determine the strength of a permanganate solution by means of an oxalic acid solution of known strength : conversely we can determine an unknown quantity of oxalic acid by means of a solution of permanganate, the strength of which is accurately known to us. Upon these considerations is based a method of determining lime volumetrically.

Weigh out about 2 grams of marble into a 250 c.c. flask, dissolve it in dilute hydrochloric acid, heat to boiling, add dilute ammonia in slight excess, and solution of ammonium oxalate so long as a precipitate is formed. Allow the precipitate to settle, pour the supernatant fluid through a small filter, and thoroughly wash the calcium oxalate by decantation with hot water. Pour the turbid liquid through the filter, but do not bring more of the precipitate on the filter than you can help. Wash the filter thoroughly, place the funnel in the neck of the $\frac{1}{2}$-litre flask, and pour into it a quantity of dilute sulphuric acid, previously heated. Again wash the filter until the filtrate—which should of course drop into the flask—is no longer acid. Add a further quantity of dilute sulphuric acid to the flask, dilute with water, and heat gently. Allow the liquid to cool, fill up the flask to the con- taining-mark, agitate, and quickly transfer 100 c.c. of the turbid liquid to a beaker, heat to about 60°, and add the permanganate solution until the pink colour is permanent. Again shake the liquid in the flask, and repeat the deter- mination on a second quantity of 100 c.c. The results should agree. 1 eq. of oxalic acid is equal to 1 eq. of calcium. The details of the calculation need no explanation.

This method is especially convenient when a number of estimations of lime have to be made in succession. It may be accelerated by adding an excess of ammonium oxalate of known strength to the solution of the lime salt, diluting to a definite bulk, allowing the precipitate to settle, withdrawing an aliquot portion of the clear liquid, acidulating with sulphuric acid, heating to 60°, and adding the permanganate until the coloration shows that the reaction is finished. We of course know the amount of ammonium oxalate we have used originally; the titration tells us the excess remaining in solution; the difference expresses that combined in the insoluble lime salt. Since 1 eq. oxalic acid is equal to 1 eq. of calcium, we can readily calculate the quantity of the alkaline earth from the amount of permanganate solution used.

XV. Volumetric Estimation of Lead by Perman- ganate Solution.

In certain cases lead may be estimated by the foregoing methods, but on account of the slight solubility of the lead oxalate in water containing ammonia, the results are not quite so accurate as in the case of lime.

XVI.   Valuation of Manganese Ores by means of
Potassium Permanganate Solution.

(See Part IV.)

XVII.   Estimation of Potassium Ferrocyanide by
Permanganate Solution.

Potassium permanganate solution added to a solution of
potassium ferrocyanide, acidulated with sulphuric acid,
converts that salt into potassium ferricyanide.   If the solution is sufficiently dilute there is no difficulty in perceiving
the termination of the reaction.   It is advisable to titrate the
permanganate solution (which should contain about 1 gram
$KMnO_4$ per 1000 c.c.) with a solution of potassium ferrocyanide of known strength; 5 grams of the recrystallised
salt ($K_4FeCy_6 + 3H_2O$) dissolved in 500 c.c. of water forms
a convenient solution.   A measured quantity of the solution,
say 25 c.c., is placed in a porcelain basin and diluted with
about ten times its bulk of water, together with a quantity
of pure sulphuric acid, and the potassium permanganate is
added from a burette, with constant stirring, until the pure
yellow colour of the solution changes to a reddish-yellow
tint.   To determine the amount of pure ferrocyanide in a
sample of the article, weigh off about 3 grams, dissolve in
water, and make up the solution to 250 c.c.   25 c.c. of the
liquid are transferred to a porcelain basin, diluted with
water, and treated in the manner directed.

XVIII.   Estimation of Potassium Ferricyanide by
Permanganate Solution.

This substance may be analysed by reducing it to the
state of ferrocyanide, and determining the amount of the
reduced salt in the manner already described.   The weighed
quantity of the ferricyanide is rendered strongly alkaline with
caustic potash solution, heated to boiling, and mixed with a
concentrated solution of ferrous sulphate, added little by

little until the colour of the precipitate is black, owing to
the formation of triferric tetroxide. The alkaline solution
is diluted, filtered, and the filtrate made up to 300 c.c.; 50
or 100 c.c. are then transferred to a porcelain basin,
strongly acidified with sulphuric acid, and titrated with
potassium permanganate solution.

B. ANALYSES BY MEANS OF IODINE AND SODIUM THIOSULPHATE (HYPOSULPHITE) SOLUTIONS.

When iodine is brought into contact with sodium thiosulphate (hyposulphite) the following reaction occurs—

$$2Na_2S_2O_3 + I_2 = Na_2S_4O_6 + 2NaI$$

iodine and sodium thiosulphate forming sodium tetrathionate
and sodium iodide. If now we mix with the thiosulphate
solution a small quantity of starch, and add the iodine solution drop by drop, the sensitive blue colour of the iodide of
starch will continue to be destroyed as fast as it is produced,
so long as the foregoing reaction occurs. Immediately that
the whole of the thiosulphate has been converted into
tetrathionate, the least excess of iodine will act upon the
starch, and the blue colour will be permanent. If, therefore,
we have a solution of thiosulphate of known strength we can
readily estimate the amount of iodine in a solution containing
an unknown quantity of that element.

A solution of iodine, in presence of substances which
readily take up oxygen, decomposes water, forming hydriodic
acid with its hydrogen and giving up the oxygen to the
oxidisable substance. Thus in the case of arsenious acid—

$$As_2O_3 + 2I_2 + 2H_2O = 4HI + As_2O_5.$$

In the case of sulphur dioxide—

$$SO_2 + I_2 + 2H_2O = SO_4H_2 + 2HI.$$

These reactions afford the basis of an exact and generally
applicable volumetric process.

The method requires—

    (1) A solution of iodine of known strength.

    (2) A solution of sodium thiosulphate of known strength.

    (3) A solution of starch.

1. *Preparation of the Iodine Solution.*—This may most conveniently be of deci-normal strength : 1 c.c. should contain therefore 0·012685 gram iodine. About 13 grams of pure iodine are weighed out into a litre flask, together with about 18 grams of pure potassium iodide, the whole is dissolved in about 300 c.c. of perfectly cold water, and diluted so as to be of deci-normal strength. Thus supposing that we had weighed out exactly 13 grams of iodine, we should require to dilute the liquid to 1024 c.c., since

$$12·685 \; : \; 1000 \; :: \; 13 \; : \; x \quad x = 1024.$$

The pure iodine may be obtained by intimately mixing resublimed iodine with about one-fourth of its weight of powdered potassium iodide, heating the mixture in a large porcelain crucible placed on an iron plate, and surmounted by a precisely similar crucible. The resublimed crystals are loosened from the sides of the crucible and placed over strong oil of vitriol within a bell-jar for a few hours, in order to deprive them of the last traces of hygroscopic moisture. The dried iodine is then quickly transferred to a clean dry test-tube, into which a second test-tube is fitted, in the

Fig. 49.   Fig. 50.

manner seen in fig. 49. The stoppered tube represented in fig. 50 may also be employed.   The tubes containing the iodine are accurately weighed, a portion of the substance is transferred to the litre-flask containing the potassium iodide, dissolved in water, and the tubes are again weighed.   The loss of weight gives the amount of iodine taken : care should be taken that the amount is not less than 12·685 grams.   Any powder adhering to the sides of the flask is washed down into the liquid ; when the whole of the iodine is dissolved the solution is diluted to the proper degree. The liquid must not be heated ; if the weighing out and the dilution have been carefully conducted the solution will be strictly deci-normal. The solution is most conveniently preserved in small stoppered bottles of about 200 c.c. capacity, which should be filled to the neck and kept in a cool, dark place.

2. *Preparation of Deci-normal Solution of Sodium Thiosulphate.*—1 litre contains 24·8 grams of the salt. About 25 grams of the recrystallised salt, previously powdered and dried between filter-paper, are accurately weighed out into the litre flask, dissolved in water, and the solution diluted so that 1 c.c. = 0·0248 gram of thiosulphate. The solution should also be kept in the dark : when exposed to light, it slowly decomposes, with the precipitation of sulphur. Accordingly a fresh solution of the salt should be prepared from time to time.

3. *Preparation of Starch Solution.*—1 gram of pure wheaten starch is mixed with a small quantity of water and rubbed to a thin cream in a mortar. The paste is poured into 150 c.c. of boiling water, the liquid is allowed to stand, and the clear solution decanted from the sediment. The solution should be prepared before the beginning of each series of experiments, since it decomposes after a time. By adding about 10 c.c. of glycerine to it, or saturating it with common salt, the solution keeps better : it is so readily prepared, however, that it is preferable to make a fresh solution when wanted.

If any doubt should exist as to the exact strength of the thiosulphate solution, it may be readily standardised by the aid of a deci-normal solution of potassium bichromate. Potassium bichromate solution added to potassium iodide, in presence of free hydrochloric acid liberates iodine :

$$K_2Cr_2O_7 + 6KI + 14HCl = 3I_2 + 8KCl + Cr_2Cl_6 + 7H_2O.$$

294·3 parts of potassium bichromate liberate 761·1 parts of iodine : accordingly 1 c.c. of deci-normal solution of bichromate liberates 0·012685 gram of iodine.

25 c.c. of deci-normal bichromate solution made by dissolving 4·907 grams of potassium bichromate in a litre of water are placed in a flask and mixed with about 10 c.c. of solution of pure potassium iodide (1 of salt to 10 of water), together with about 5 c.c. of pure hydrochloric acid, and 200 c.c. of water. The standard thiosulphate solution is then added drop by drop until the iodine has nearly disappeared : a few drops of starch solution are added, and the addition of the thiosulphate continued until the last trace of the blue colour of the iodide of starch vanishes. Since the 25 c.c. of bichromate solution liberate 25 × 0·012685 = 0·3171 gram iodine, we can readily calculate the amount of iodine equivalent to 1 c.c. of the sodium thiosulphate solution.

XIX. Valuation of Bleaching-powder by Iodine and Sodium Thiosulphate Solutions.
(See Part IV.)

XX. Estimation of the Amount of Chlorine in Aqueous Solutions of the Gas.

Prepare a dilute solution of chlorine in water, measure off a definite quantity of the liquid, and transfer it to a solution of potassium iodide. Determine the amount of the liberated iodine by solution of sodium thiosulphate, adding the latter liquid until the iodine has nearly disappeared ; add 2 or 3 c.c. of starch solution, and continue the addition of the thiosulphate until the blue colour just vanishes.

Add a second portion of the chlorine water to the solution of potassium iodide, and a measured quantity (in excess) of sodium thiosulphate, and determine the amount of the residual thiosulphate by means of starch and standard iodine solution, adding the latter until the blue colour of the solution is persistent. The results of the two experiments should agree.

XXI. Estimation of the Amount of Sulphur Dioxide in Aqueous Solutions of the Gas.

Prepare a very dilute solution of sulphur dioxide (by adding 10 c.c. of a saturated solution of the gas to a litre of water), transfer a definite quantity of the liquid to a known amount (in excess) of standard iodine solution, and determine the amount of residual iodine by starch and sodium thiosulphate solution.

To a second measured portion of the solution of the gas, add 2 or 3 c.c. of starch solution, and the standard solution of iodine, until the blue colour of the iodide of starch is persistent. The reaction between sulphur dioxide and iodine may be thus represented :

$$SO_2 + I_2 + 2H_2O = SO_4H_2 + 2HI.$$

If, however, the solution is concentrated, the sulphuric acid reacts upon the hydriodic acid. and sulphur dioxide and free iodine are again formed—

$$SO_4H_2 + 2HI = SO_2 + I_2 + 2H_2O.$$

So long as the solution contains only about 0·05 per cent. of sulphur dioxide, the first reaction alone takes place.

XXII. Estimation of Sulphuretted Hydrogen in Aqueous Solutions of the Gas.

When sulphuretted hydrogen is brought into contact with free iodine, the following reaction ensues :—

$$H_2S + I_2 = 2HI + S.$$

This reaction, however, is liable to be modified by the concentration of the solution ; experiment has shown that it can only be depended upon when the solution contains not more than 0·04 per cent. of the gas.

Prepare a dilute solution of sulphuretted hydrogen, measure off a definite quantity, add starch liquor and solution of iodine until the colour is persistent.

Place in a flask about the same quantity of iodine solution which you have consumed in the foregoing experiment, and then add a measured quantity (in excess) of the solution of sulphuretted hydrogen. Determine the excess of sulphuretted hydrogen by starch and standard iodine solutions.

The amount of sulphuretted hydrogen in mineral waters may be readily determined by this method. For such estimations it will be more convenient to dilute the iodine solution to ten times its bulk ; 100 c.c. are transferred to the litre flask, and the flask is filled up to the containing-mark.

Measure off 250 c.c. of the mineral water, transfer to a beaker, add 1 or 2 c.c. of starch liquor and centi-normal solution of iodine, until the blue colour is persistent. Let us suppose we have used 20 c.c. of centi-normal solution of iodine ; this amount of the iodine solution is placed in a beaker, and mixed with 250 c.c. of the mineral water ; 1 or 2 c.c. of starch liquor are added, and then, cautiously, centi-normal solution of iodine, drop by drop, until the blue colour of the iodide of starch remains. It will be found in general that 1 or 2 c.c. more of iodine solution are required in the second experiment than in the first. Now pour 250 c.c. of distilled water into the beaker, add the same bulk of starch liquor used in the last experiment, and, drop by drop, the solution of iodine, until the blue colour is properly defined. This experiment gives the amount of the iodine solution required to produce the final reaction; subtract this quantity from the amount of iodine solution required in the second experiment. The remainder shows the amount of iodine solution equivalent to the sulphuretted hydrogen present ; this amount may readily be calculated from the above equation.

XXIII. ESTIMATION OF HYDROCYANIC ACID.

When potassium cyanide is mixed with solution of iodine, iodide of potassium and iodide of cyanogen are formed :—

$$KCy + I_2 = KI + CyI.$$

Two eq. or 253·7 parts of iodine correspond to 1 eq. or 65·17 of potassium cyanide. This principle affords the basis of an exact method for determining the value of potassium cyanide, or solutions of prussic acid.

In the case of potassium cyanide, weigh out about 2 grams of the salt into a ½-litre flask, dissolve in water, dilute to the mark, shake, and transfer 50 c.c. to a beaker containing about 200 c.c. of water, add 100 c.c. of a saturated solution of carbonic acid gas in water (to convert the alkaline carbonates, always present in the commercial article, to acid carbonates), and add solution of iodine, until the liquid possesses a slight and permanent yellow tinge. The cyanide must be free from alkaline sulphide.

In the case of free hydrocyanic acid, add a very slight excess of caustic soda, and then solution of carbonic acid, and proceed in the manner above described. The specific gravity of the ordinary preparations of hydrocyanic acid is so little removed from that of water, that it is more prudent to determine the amount taken for analysis by weighing, rather than by measuring the solution in a pipette, by aspiration in the ordinary manner.

XXIV. ESTIMATION OF ANTIMONY IN TARTAR EMETIC, AND OF ARSENIC AND ARSENIOUS ACIDS IN COMMERCIAL ARSENIATES.

These methods depend upon the conversion of antimony or arsenic trioxide, in an alkaline solution, into pentoxide, by solution of iodine:

$$Sb_2O_3 + 2I_2 + 2H_2O = Sb_2O_5 + 4HI$$
$$As_2O_3 + 2I_2 + 2H_2O = As_2O_5 + 4HI$$

Weigh out about 2 grams of the tartar emetic, dissolve in water, and dilute to 250 c.c. Transfer 20 c.c. of the solution to a beaker, add the same amount of a saturated solution of sodium bicarbonate, and 2 c.c. of starch liquor. Now add the iodine solution until the *blue* colour is

M

persistent for about 5 minutes. The fluid acquires a reddish tint just before the reaction is completed : the blue colouration, even after the reaction is finished, fades after a time, say in 15 minutes.

In the case of commercial arseniates weigh out about 3 grams of the substance into a $\frac{1}{2}$-litre flask, and dissolve in about 150 c.c. of warm water, add a small quantity of sodium acetate and acetic acid, and boil the solution for 15 minutes to decompose any nitrites : when cold make up to 500 c.c.

To determine the amount of the arsenite withdraw 50 c.c. of the solution, add 25 cb.c. of a saturated solution of pure $NaHCO_3$, starch paste, and decinormal iodine solution (1 c.c. = ·00495 gram As_2O_3).

Now pass sulphur dioxide gas into the solution in the flask to reduce the pentoxide

$$As_2O_5 + 2SO_2 + 2H_2O = As_2O_3 + 2H_2SO_4,$$

boil to expel the excess of sulphur dioxide, and when cold make up to 500 c.c. Withdraw 50 c.c. of the solution, add 25 c.c. of the sodium bicarbonate solution, starch paste and iodine solution as before (1 c.c. =0·00575 gram As_2O_5).

XXV. Determination of Tin by Iodine Solution.

The metal is dissolved in hydrochloric acid, and mixed with a solution of Rochelle salt, and a concentrated solution of sodium bicarbonate is then added until the liquid is no longer acid. Add about 1 c.c. of starch solution and decinormal solution of iodine, until the blue colour is persistent. 253·7 parts of iodine are equivalent to 118 of tin. The solution of the metal is best effected in a stream of carbon dioxide : the addition of a few scraps of platinum-foil accelerates the process.

This method is of course applicable to the valuation of 'Tin crystals.'

ANALYSES BY MEANS OF IODINE AND SODIUM THIOSULPHATE SOLUTIONS, WITH PREVIOUS DISTILLATION WITH HYDROCHLORIC ACID.

A number of substances containing oxygen, when heated with hydrochloric acid, are decomposed in such manner that free chlorine is evolved in amount bearing some simple ratio to the quantity of oxygen present. Thus when manganese dioxide is heated with hydrochloric acid, the following reaction occurs:

$$MnO_2 + 4HCl = MnCl_2 + Cl_2 + 2H_2O$$

70·92 parts of chlorine are therefore equivalent to 86·04 of manganese dioxide. If the chlorine be led into a solution of potassium iodide, iodine will be liberated in exact proportion to the chlorine evolved: the amount of iodine liberated is a measure therefore of the amount of real manganese dioxide in the sample analysed.

Similarly when potassium bichromate is heated with hydrochloric acid, chlorine is evolved:

$$K_2Cr_2O_7 + 14HCl = Cr_2Cl_6 + 7H_2O + 2KCl + 3Cl_2.$$

212·76 parts of chlorine are equivalent therefore to 294·3 parts of potassium bichromate. As we have seen from the method of titration given on p. 158, No. XX. the disengaged chlorine, in presence of excess of potassium iodide, liberates an equivalent quantity of iodine, which can be estimated by means of starch and sodium thiosulphate solutions.

XXVI. ANALYSIS OF POTASSIUM BICHROMATE.

Into the little bulb *a*, which has a capacity of about 60 c.c. (fig. 50A), accurately weigh out about 0·3 or 0·4 gram of pure fused potassium bichromate, add about 25 c.c. of pure fuming hydrochloric acid, and connect the bulb with the bent tube *b* by means of a tightly-fitting caoutchouc tube, which has been previously boiled in caustic soda solution to remove any adhering sulphur. The bent tube is connected with a two-bulb U-tube by means of a caoutchouc cork, which should also have been previously cleansed by caustic soda solution. In certain cases it is advisable to have a second U-tube, which is connected with the first by corks and a bent tube. Both the tubes are placed in a beaker and are surrounded by cold water: they each contain about 25 c.c. of strong solution of potassium iodide. Gently heat the bulb containing the bichromate and acid ; chlorine is readily evolved

FIG. 50A.

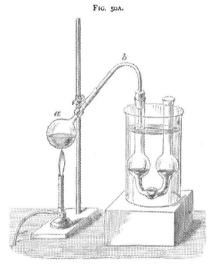

and decomposes the potassium iodide in the U-tubes: after two or three minutes' heating, the whole of the chlorine will have been eliminated. Heat the liquid to boiling, so as to drive over the last traces of the gas ; remove the lamp, and allow the whole to stand for 5 or 10 minutes to effect the complete absorption of the chlorine ; empty the contents of the U-tubes, when quite cold, into a beaker (the solution in the second tube need not be added unless it contains liberated iodine), dilute with water, and titrate with starch and thiosulphate solution.

XXVII. ESTIMATION OF ARSENIOUS ACID BY THE AID OF THE FOREGOING REACTION.

If we mix the weighed quantity of bichromate with finely-powdered arsenious acid (not in excess) only a portion of the chlorine demanded by the equation is evolved; the deficit has served to bring about the oxidation of the arsenic trioxide to pentoxide:

$$As_2O_3 + 4Cl + 2H_2O = As_2O_5 + 4HCl.$$

Consequently every 4 eq. of missing chlorine, i.e. less than that which would be obtained by distilling the weighed amount of bichromate alone with hydrochloric acid, represent 1 eq. of arsenic trioxide. The details of the method are identical with those of the foregoing example.

XXVIII. ANALYSIS OF CHLORATES, BROMATES, AND IODATES.

1 eq. of potassium chlorate, heated with a large excess of hydrochloric acid, evolves an amount of chlorine partly free and partly in combination with oxygen sufficient to liberate 6 eq. of iodine. 761·1 parts of iodine are therefore equivalent to 122·56 of potassium chlorate. The weighed quantity of the chlorate (about 0·2 gram is sufficient) is placed in the distillation flask and heated with excess of hydrochloric acid. The remainder of the operation is conducted in the manner described. Bromates and iodates are best analysed by the method of digestion instead of by that of distillation. A strong bottle of about 120–150 c.c. is fitted with an accurately-ground stopper: the weighed amount of the bromate is placed in the bottle, the requisite amount of a saturated solution of potassium iodide and hydrochloric acid is added, and the stopper is firmly fastened down by binding wire. The bottle is then placed on the water-bath; when the decomposition is complete it is allowed to cool, and its contents are diluted with water; the solution is emptied into a beaker, and the titration proceeded with in the usual manner.

In the case of iodates and bromates, only 4 eq. of iodine are liberated for each eq. of the acid.

The method of digestion may be frequently substituted for that of distillation. It is of course absolutely necessary that the stopper of the bottle fits perfectly : to test it, it should be tied down in the empty bottle, which is then to be immersed in hot water ; if any air-bubbles make their escape between the stopper and the neck, the bottle is useless for this purpose. The stopper, if nearly tight, may be re-ground with a little fine emery and water. In every case it must be carefully tested before use.

XXIX. Estimation of Iron by means of Iodine and Thiosulphate Solutions.

When ferric chloride is added to a warm solution of potassium iodide the following reaction ensues :

$$Fe_2Cl_6 + 2KI = 2FeCl_2 + 2KCl + I_2.$$

126·85 parts of iodine correspond to 56 parts of iron.

Dissolve 5·02 grams of clean piano-wire (corresponding to 5 grams of pure iron) in dilute hydrochloric acid, in a $\frac{1}{2}$-litre flask ; add a few crystals of potassium chlorate, to convert the iron into ferric chloride, boil the solution for some time, to expel the excess of chlorine, and when cold dilute the solution to the containing-mark. 10 c.c. of the solution correspond to 0·1 gram of iron. Transfer 20 c.c. of the iron solution to the stoppered bottle, cautiously add dilute caustic soda solution until a slight precipitate of oxide of iron remains, and then 1 c.c. of hydrochloric acid (sp. gr. 1·1). Now add about 4 grams of potassium iodide to the clear dark-yellow solution, insert the stopper, and fasten it down by means of binding wire. Heat the bottle over the water-bath, for ten or fifteen minutes, to about 60°, allow it to cool completely, open it, and add sodium thiosulphate from a burette, until the solution is nearly decolourised, add 1 c.c. of starch liquor, and continue the addition of the thiosulphate until the blue colouration just disappears.

This method is particularly useful for the determination of small quantities of iron. The solution must, of course, be fully oxidised, and contain no other substance which can eliminate iodine. If strongly acid, it must be partially neutralised by soda solution, in the manner described.

XXX. ESTIMATION OF NITRIC ACID BY SOLUTIONS OF IRON, IODINE, AND SODIUM THIOSULPHATE.

Free nitric acid, added to a solution of ferrous chloride, converts the iron into ferric chloride. Ferric chloride, as we have seen in the foregoing process, may be estimated by means of iodide of potassium and sodium thiosulphate.

Weigh out about 0.7 gram of iron-wire, and dissolve it in a small quantity of hydrochloric acid, in a flask in a current of carbon dioxide, in order to prevent the least chance of oxidation. Weigh out about 2 grams of nitre into a ¼-litre flask, dissolve in boiling water, and dilute with boiled water to the containing-mark. Transfer 25 c.c. of the solution, corresponding to 0.2 gram of nitre, to the iron solution, and quickly re-insert the cork ; gently heat the liquid, at length to boiling, to expel the nitric oxide. Maintain a rapid current of carbon dioxide throughout the operation. As soon as the solution is of a pure yellow colour, allow it to cool in a current of carbon dioxide, add a sufficiency of potassium iodide, allow the solution to stand for a short time, and determine the amount of liberated iodine by starch and thiosulphate solution, exactly as described in the preceding method. The quantity of thiosulphate used, multiplied by 0.0021, gives the amount of nitric acid present.

XXX. VALUATION OF MANGANESE ORES BY DISTILLATION WITH HYDROCHLORIC ACID, AND TITRATION WITH IODINE AND THIOSULPHATE SOLUTIONS.

(See Part IV.)

PART IV.

GENERAL ANALYSIS, INVOLVING GRAVIMETRIC AND VOLUMETRIC PROCESSES.

I. NITRE.

THE crude nitre of commerce invariably contains alkaline chlorides and sulphates, together with more or less insoluble matter and moisture. Samples of nitre are occasionally met with containing sodium nitrate, arising either from imperfect decomposition of the Chili saltpetre, from which the nitre was prepared, or from wilful adulteration. Such nitre is highly hygroscopic, and requires to be purified before it can be used for the manufacture of gunpowder.

The presence of sodium nitrate, or excess oi common salt in the nitre, may be readily detected by means of the spectroscope. The quantity of admixed sodium nitrate may be approximately determined by ascertaining the amount of water taken up from a perfectly moist atmosphere. Pure nitre placed over the surface of water for a fortnight remains comparatively dry ; sodium nitrate, under the same circumstances, absorbs one-fourth of its weight of water. The following table shows the amount of water taken up by 100 grams of the mixed nitrates :—

Percentage of Sodium Nitrate	0·5	1	3	5	10
Amount of Water (in grams) absorbed in 14 days	2·5	4	10	12	19

Determination of Moisture.—About 20 grams of the sample are gently heated in a weighed platinum crucible, until the salt commences to fuse. When cold, the crucible is again weighed. The loss gives the amount of moisture.

Determination of Insoluble Matter.—The contents of the crucible are washed out into a porcelain basin with hot water, dissolved, filtered, and the insoluble matter dried and weighed. The filtrate is received into a ½-litre flask, allowed to cool, and diluted to the containing-mark.

Determination of Chlorine.—100 c.c. of the liquid, corresponding to 4 grams of nitre, are transferred to a porcelain basin, and the chlorine estimated by means of standard silver nitrate and potassium chromate solutions. If the amount of chlorine is very small, it is advisable to use centinormal silver solution.

Determination of Sulphuric Acid.—250 c.c. of the liquid are transferred to a beaker, heated to boiling, and acidulated with a small quantity of hydrochloric acid. Barium chloride solution is added, and the liquid is set aside for a time, to allow the precipitate to subside perfectly. As barium sulphate is slightly soluble in solutions of the alkaline nitrates, the precipitation is not quite complete. The clear supernatant liquid is poured through the filter, and the precipitate is washed several times by decantation with boiling water, taking care to allow it to settle as completely as possible, before pouring the liquid on to the filter. The barium sulphate still retains co-precipitated nitrate. This may be best removed by a solution of copper acetate. Crystallised copper acetate is dissolved in a small quantity of hot water, containing acetic acid ; one drop of sulphuric acid is added, and then one drop of barium chloride solution ; the mixture is boiled and filtered. About 5 or 10 c.c. of the saturated solution, according to the amount of the precipitate, is added to the barium sulphate, together with a small quantity of water and a few drops of acetic acid. The liquid is heated to boiling, and maintained in ebullition for ten or fifteen minutes. The amount of acetic acid added should be sufficient to prevent the precipitation of any basic salt of copper. Pour the liquid through the filter, transfer the precipitate, and wash it thoroughly with hot water. Dry, ignite, and weigh it. This procedure is recommended to be followed in all precipitations of sulphuric acid by barium salts, in presence of considerable quantities of alkaline nitrates or chlorides.

Determination of Nitric Acid.—Fuse a few grams of the nitre at the lowest possible temperature, and pour out the liquid mass into a warm porcelain dish, powder it quickly, and transfer it to a tube. Place 2 or 3 grams of powdered quartz in a platinum crucible, heat to redness, and weigh accurately after cooling. Add about 0·5 gram of the nitre, and again weigh. Mix the nitre and silica by the aid of a thin glass rod, taking care, of course, that nothing adheres to the rod, and heat the crucible gradually to a low red heat, keeping it at this temperature for twenty or thirty minutes, transfer to the desiccator, and weigh when cold. The loss of weight gives the amount of nitric acid. Care must be taken to regulate the temperature properly, or the sulphates and chlorides present (particularly the latter) may partially volatilise. Potassium bichromate and borax may be substituted for the powdered quartz ; the latter substance, however, is preferable.

The nitric acid may also be determined by the method described on p. 96.

II. GUNPOWDER.

Gunpowder is an intimate mixture of sulphur, nitre, and charcoal. It invariably contains also a small quantity of moisture. In the following scheme of analysis all these constituents are determined in a single portion of the sample.

A light glass tube (*a*, fig. 51), about 10 centimetres long

and 1 centimetre in diameter, is drawn out near the end by
the aid of the blow-pipe. The contracted portion should
measure about 5 centimetres long, and possess an internal
diameter of 0·2 centimetre. At the point where the tube is
narrowed, place a plug of recently-ignited asbestos, from 1·5
to 2 centimetres long. Accurately weigh the tube, place in
it about 3 grams of the triturated powder, and again
weigh: the increase of course gives the amount taken for

FIG. 51.

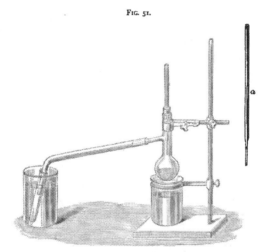

analysis. By the aid of the filter-pump aspirate a gentle current of air, dried by sulphuric acid, through the tube for 10
or 12 hours, and again weigh. The loss indicates the amount
of *moisture.*

The tube is next fitted into a light flask of about 25 c.c.
capacity, provided with a side-tube in the manner seen in
fig. 51. The flask is accurately weighed and connected with
a tube of thin glass 25 to 30 centimetres long. Portions of

about 3 c.c. of carbon bisulphide (free from moisture, and
rectified by agitation with mercury, and redistillation) are
poured over the powder. The filtrate running into the
flask should be perfectly clear. As soon as the flask is
about half filled, it is heated by hot water at about 70°.
The carbon bisulphide distils over, and is collected in a
dry test-tube surrounded by cold water. The distillate is
again poured on to the powder, and again distilled, the
operation being repeated 6 or 8 times. After the last distillation the residual sulphur is gently heated, and a current of
dry air is drawn through the entire apparatus by attaching
the side-tube to the filter-pump. The flask is re-weighed :
the increase in weight gives the amount of the *sulphur* in the
powder which it is possible to extract by means of bisulphide
of carbon.

The tube containing the residual charcoal and nitre,
together with the minute quantity of sulphur still left in the
exhausted powder, is heated to 100°, and a current of air,
dried by sulphuric acid, is drawn over it. The tube is again
weighed : its decrease in weight will be slightly greater than
the increase in weight of the flask containing the sulphur.
The difference is the amount of *moisture* which the powder
would give up if dried at 100°. This slight difference is to
be added to the quantity of moisture already determined.

About 1 gram of the exhausted powder is shaken out into
a porcelain basin, and the tube is re-weighed. The loss of
weight shows the amount transferred to the basin. This
powder is then gently heated with nitric acid (perfectly free
from sulphuric acid) and a few crystals of potassium
chlorate. The liquid is evaporated to dryness with hydrochloric acid, dissolved in water, filtered if necessary, and
the sulphuric acid precipitated in the ordinary manner by
barium chloride. The barium sulphate is dried, ignited, and
weighed, and the amount of sulphur equivalent to it is added
to the main quantity extracted by the sulphide of carbon.

The tube containing the remainder of the exhausted powder is now treated with water to extract the nitre. It is fitted into the bell-jar (fig. 52), standing on a plate of ground glass. Within the jar and underneath the tube is a weighed platinum dish. The side tube is connected with the filter-pump. A few cubic centimetres of cold water are poured on to the powder, and the pump is set in operation so as to cause the liquid to fall, drop by drop, into the basin. To avoid loss by splashing,

Fig. 52.

the basin should be as near to the edge of the tube as possible. Successive small quantities of water of a gradually increasing temperature are now poured over the powder, water as hot as possible being used for the last washings. 50 c.c. of water are amply sufficient to extract all the nitre from the residue of the powder, if care be taken not to use too much water for each washing. The use of large quantities of water for washing is not advisable, since appreciable quantities of organic matter are thereby liable to be dissolved out of the charcoal. The solution of the nitre is evaporated to dryness in the dish, dried at 120°, and weighed: the weight is of course calculated upon the entire quantity of powder. The asbestos-plug is now detached from the tube by the aid of a platinum wire, and the tube and its contents are again dried at 100°, and weighed. It will be generally found, if the process has been properly carried out, that the weight of the charcoal is slightly greater (generally from 1 to 2 milligrams) than the amount calculated from the quantity of nitre found: the difference is due to the fact that the pure charcoal retains water more tenaciously, even after drying for some time at 100°, than when mixed with nitre.

It is sometimes necessary to determine the amount of carbon, hydrogen, and oxygen in the residual charcoal : this may readily be effected by mixing it, together with the asbestos, with lead chromate, and burning it in a stream of oxygen, and collecting the carbon dioxide and water in the manner described under organic analysis. In calculating the amount of hydrogen, due regard must be taken of the water retained by the charcoal after drying at 100°.

III. Limestones. Hydraulic Mortar.

Limestone is essentially calcium carbonate, containing more or less magnesium carbonate, ferrous and manganous carbonates or oxides, alumina, silica, and alkalies. Many limestones also contain variable quantities of clay, sand, and organic matter, together with chlorine, fluorine, phosphoric and sulphuric acids, and iron pyrites.

The method given on p. 85, which includes the determination of the essential constituents, will generally suffice for the examination of limestone for technical purposes, but occasionally it is required to estimate the substances which are present in smaller proportion.

Powder about 100 grams of the mineral, mix uniformly, and dry at 100°.   Weigh out about 2 grams into the flask A (fig. 31), and determine the amount of carbon dioxide in the manner directed on p. 86.   Rinse the solution into a porcelain basin, evaporate to complete dryness, moisten with a few drops of hydrochloric acid, dissolve in hot water, filter through a weighed filter, wash the residue, and dry it at 100°. It may consist of sand, clay, and separated silicic acid, and organic matter.   The proportion of these several substances will be estimated hereafter.

Add a few drops of bromine-water to the filtrate, and then ammonium chloride and a *slight* excess of ammonia, cover the beaker, and heat it gently for some time.   The precipitate contains the oxides of iron, manganese, and aluminium, together with the phosphoric acid ; it is thrown on to a filter, washed once or twice, re-dissolved in a small quantity of hydrochloric acid, again mixed with bromine-water, and re-precipitated with ammonia, and again filtered. The second precipitation effects the removal of the small quantities of lime and magnesia which are invariably thrown down with the oxides on the first precipitation. The precipitate is well washed, dried, and weighed.

The lime and magnesia in the mixed filtrate are separated as directed on p. 88. The lime may be estimated volumetrically, as described on p. 152, or it may be weighed as carbonate or oxide.

Determination of the Constituents present in small quantity.—Dissolve about 50 grams of the mineral in dilute hydrochloric acid in a porcelain basin, heat the solution gently to expel carbon dioxide, and filter through a weighed filter into a litre flask, wash the residue thoroughly, dry, and weigh it. Dilute the filtrate up to the containing-mark.

Analysis of the Insoluble Residue.—(*a*) Weigh out about one-fourth of the insoluble matter into a platinum basin, and boil it with strong solution of pure sodium carbonate. Filter, and determine the silicic acid in solution by acidulation with hydrochloric acid, and evaporation to dryness in a platinum basin.

(*b*) Weigh out another portion into a platinum crucible and fuse with pure sodium and potassium carbonates, extract with hot water, acidulate with hydrochloric acid, evaporate to dryness, and separate the silica in the usual manner. Deduct the amount of the silica soluble in solutions of alkaline carbonate (*a*).

(*c*) A third portion of the residue is weighed out into a platinum boat and heated in a current of oxygen in a combustion-tube partly filled with copper oxide, in order to determine the amount of organic matter (humus). According to Petzholdt humus contains 58 per cent. of carbon.

(*d*)   Iron pyrites is not an infrequent constituent of limestones : it is found in the insoluble residue, after treatment with dilute hydrochloric acid.   To determine its amount the remainder of the insoluble residue is heated with nitric acid and potassium chlorate, and the proportion of the pyrites is calculated from the quantity of sulphuric acid obtained.

Analysis of the Solution.—Transfer 500 c.c. of the liquid to a porcelain basin, evaporate to complete dryness, and heat the saline mass until fumes of hydrochloric acid are no longer visible.   Moisten with strong hydrochloric acid, add hot water, and filter the solution, wash the residue, and weigh it in a platinum crucible.   It consists mainly of silica, but may contain sulphates of strontium and barium.   Call it Pp. I.

To the filtrate add a few drops of nitric acid, boil, add ammonia, and again boil until the fluid no longer smells of ammonia, filter, wash the precipitate once or twice, dissolve it in hydrochloric acid, and precipitate again with ammonia. Call it Pp. II. : it contains ferric oxide, alumina, and phosphoric acid.   The mixed filtrates are received in a flask (which they should nearly fill), and mixed with ammonium sulphide. The flask is closed and set aside for 24 hours in a warm place.   The precipitate consists of manganese sulphide. Filter it off and wash it with water containing ammonium sulphide.   Call it Pp. III.

Treatment of the Precipitates I., II., and III.—Pp. I. Moisten the weighed precipitate with pure hydrofluoric acid and a drop or two of sulphuric acid, and evaporate to dryness.   Repeat this operation, and fuse any residue with sodium carbonate, digest with hot water, filter, dissolve the washed precipitate in hydrochloric acid, and add a drop or two of sulphuric acid to the solution.   Filter into a small flask (filtrate *a*), wash the precipitate, and digest it on the filter for 12 or 15 hours with solution of ammonium carbonate, the tube of the funnel being meanwhile closed by a rod during the digestion. Open the tube, wash the precipitate with water, and treat it with dilute hydrochloric acid (filtrate *b*), again wash with water, and weigh the residual barium sulphate. Mix *a* and *b*, add ammonia and ammonium carbonate; if a precipitate forms on standing it consists of strontium carbonate; it is filtered, washed with ammonia water, dried, and weighed.

Pp. II. Dissolve in hydrochloric acid in a small flask, add pure tartaric acid, ammonia, and ammonium sulphide. Close the flask, and after standing a few hours, wash the iron sulphide with water containing a few drops of ammonium sulphide. Dissolve in hydrochloric acid, add a crystal or two of potassium chlorate, boil, and precipitate with ammonia, and weigh the ferric oxide.

To the yellow-coloured filtrate, containing the alumina and phosphoric acid, add a small quantity of pure sodium carbonate and nitre, evaporate to dryness, and ignite until the residue is free from carbon. Digest with water once or twice, pour off the solution into a beaker, and treat the residue with warm hydrochloric acid, add the solution to that contained in the beaker, filter if necessary, and mix the filtrate with ammonia. Filter off the precipitate and weigh it. Mix the filtrate with a few drops of magnesia mixture : if a precipitate is again formed, it consists of ammonium magnesium phosphate : it is washed with ammonia water and weighed as magnesium pyrophosphate. In that case the first precipitate has the composition $Al_2P_2O_8$: if no precipitate form, it is probably a mixture of alumina and aluminium phosphate. It is redissolved in a small quantity of hydrochloric acid and mixed with molybdic acid solution, and treated as directed under Iron Ores—Estimation of Phosphorus.

Pp. III. consists almost entirely of manganese sulphide. It is treated with moderately dilute acetic acid, and the solu-

N

tion is filtered if necessary ; if any insoluble matter remains, it is tested for the metals of Group III. The filtrate is heated to boiling, nearly neutralised with caustic soda, and mixed with bromine water, and the manganese dioxide filtered, washed, dried, ignited, and weighed as trimanganic tetroxide. *

Determination of the Alkalies.†—Transfer 300 c.c. of the solution to a flask, add bromine water, heat gently, and mix with ammonia and ammonium carbonate. Allow the liquid to stand for some hours, filter, wash, evaporate the filtrate to dryness in a platinum dish, and ignite to remove the ammonia salts. Dissolve in water, boil with a little milk of lime, filter, wash, remove the excess of lime by ammonium carbonate and oxalate, filter, wash, and evaporate the filtrate to dryness ; again ignite, dissolve in a few drops of water, filter once more if necessary, acidulate with hydrochloric acid, and evaporate the solution of the mixed alkaline chlorides to dryness in a weighed platinum dish. The potassium and sodium are then separated by means of platinum tetrachloride, as directed in No. IV. p. 85.

Determination of Sulphuric Acid.—To the remainder of the original solution, add one or two drops of barium chloride, and allow the liquid to stand for some time. If any precipitate of barium sulphate is formed, filter it off, wash, dry, and weigh it.

* Chatard determines the small quantity of manganese present in dolomites, limestones, &c., by dissolving the mineral in dilute nitric acid, boiling the solution with a small quantity of lead peroxide, filtering through an asbestos filter, and volumetrically determining the amount of permanganic acid in the liquid by a dilute solution of ammonium oxalate of known strength.

† Alkalies when present in limestones or dolomites may be readily detected by strongly heating the mineral in a platinum crucible, boiling with a little water, filtering, acidulating with hydrochloric acid, adding ammonia and ammonium carbonate, filtering, evaporating the filtrate to dryness, and examining the residue by means of the spectoscope. The ammonium carbonate precipitate may also be treated with hydrochloric acid, evaporated to dryness, and examined for barium and strontium in the same manner. (ENGELBACH.)

Determination of Chlorine.—Chlorides are occasionally present in dolomites and limestones : the amount of chlorine in them may be determined by dissolving a quantity of the mineral in dilute nitric acid, filtering, and adding a few drops of silver nitrate to the filtrate, and treating the silver chloride as directed in No. II. p. 81.

Determination of Fluorine.—A large quantity of the mineral is dissolved in acetic acid, the solution is evaporated to dryness, and heated to expel the excess of acetic acid. The mass is repeatedly treated with water, the residue is weighed, and a portion is tested for fluorine. If this substance is found in estimable quantity, fuse the remainder of the insoluble residue with sodium carbonate, boil with water, filter, and wash with boiling water and solution of ammonium carbonate. Heat the filtrate, which contains all the fluorine, with an additional quantity of ammonium carbonate, and after some time filter off the precipitated silica and alumina. Boil the filtrate until the ammonium carbonate is *completely* expelled, and add a few drops of calcium chloride solution ; if any precipitate of calcium fluoride forms, filter it off, wash thoroughly with hot water, dry, ignite, and weigh. To ensure the absence of calcium carbonate in the weighed precipitate, digest it with dilute acetic acid, allow the precipitate to subside, pass the liquid through a small filter, wash by decantation, dry the precipitate and filter, burn the latter, add the ash, and again weigh.

Determination of Water retained in the Mineral after heating to 100°.—This is effected in the apparatus seen in fig. 53. The weighed quantity of the mineral (about 3 grams) is heated in the bulb, made of difficultly-fusible glass, in a stream of aspirated air, dried by passing through the first calcium-chloride tube ; the water evolved is absorbed in the second weighed tube, which also contains calcium chloride. The increase in the weight of this tube at the termination of the experiment gives the amount of moisture present in the mineral. The little flask contains strong sulphuric acid; it serves to indicate the rate of the current of air, and prevents the possibility of moisture diffusing into the weighed calcium-chloride tube.

CEMENT-STONE, as the material used in the manufacture of hydraulic mortar is termed, is an impure limestone containing a considerable quantity of ferrous carbonate, alumina,

FIG. 53

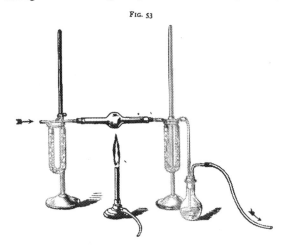

and silica. Hydraulic mortar owes its property of hardening under water to the gradual formation of hydrated silicates of lime and alumina, which are very dense and insoluble in water.

IV. CLAYS.

Clay is a hydrated aluminium silicate, derived from the decomposition of felspar. The purer varieties are perfectly white, and contain but small quantities of lime, magnesia, and

oxide of iron. The red colour of the common brick-clay
after burning is due to the ferric oxide which it contains.
The different clays used in the arts may be classed under
the heads of slate clay, common clay, fire-clay, plastic clay,
and kaolin. These varieties have essentially the same qualitative composition ; their different properties are mainly
due to the relative amounts of the admixed substances.
Pure clay is nearly infusible ; but if mixed with a sufficient
amount of iron and lime, it may be more or less readily
melted, especially if free silica be present. The ease with
which it may be fused depends upon the proportion of these
admixtures. The following analyses will serve to indicate
the composition of ordinary fire-clays :—

No. 1, Best Stourbridge Clay. No. 2, Inferior Fire-clay,
Faulty Glass Pot. No. 3, Brick Clay of average quality.

	1.	2.	3.
Silica	73·82	69·91	49·44
Alumina	15·88	17·44	34·26
Ferrous Oxide	2·95 Fe_2O_3	2·89 Fe_2O_3	7·74
Lime	trace	3·08	1·48
Magnesia	trace	4·47	5·14
Alkalies	0·90	2·21	—
Water	6·45	—	1·94
	100·00	100·00	100·00

The silica in clay exists partly as sand, partly as hydrate,
and partly in combination with bases ; the amount of free
hydrated silica seldom exceeds 1 per cent. The sand may
vary from 15 to 60 per cent.

The porcelain earth, or kaolin of the Chinese, is almost
pure hydrated silicate of aluminium, containing undecomposed felspar and free silica. It has been formed from
orthoclase by the gradual abstraction of the whole of the
alkalies, and of about ⅔ of the silica. When freed from felspar
and admixed silica, its average composition is $Al_2O_3 2SiO_2 +$
$2H_2O$, but varieties of kaolin are frequently found in which
the relation of alumina to silica is very different.

For many of its applications, it is important to know the

relative proportions of coarse and fine sand present in the sample, in addition to the true clay. The chemical examination, is, therefore, usually preceded by a mechanical analysis.

A. *Mechanical Analysis.*

This may be effected with sufficient accuracy by elutriation. The powdered clay is agitated in a stream of water; the coarse particles subside, the finer particles are carried away by the current, and are received in a large beaker, or other suitable vessel, where they are allowed to settle. Of the several forms of apparatus proposed for this purpose, the simplest is that devised by Schulze. To a tall and narrow glass (one of the old forms of champagne glass does very well) about 20 centimetres deep, and 7 centimetres in diameter at the mouth, is fastened a brass rim, about 2 centimetres broad, carrying a short tube inclined downwards. A slow stream of water is allowed to pass into the glass, through a funnel-tube about 40 centimetres long, and 7 millimetres in diameter; the bulb of the funnel should be about 5 centimetres in diameter, and its end should be drawn out until the opening is only $1\frac{1}{2}$ millimetre in diameter. Triturate 30 or 40 grams of the air-dried clay in a mortar, transfer it to a porcelain dish, and boil it with about 70 c.c. of water for thirty minutes. Repeatedly crush the sedimentary matter with a pestle, and agitate the liquid so as to disintegrate the clay completely. Allow to cool, and transfer the contents of the dish to the narrow glass; suspend the tube-funnel within the glass, so that its end is about 2 or 3 millimetres from the bottom, and so regulate the stream of water that the funnel is kept constantly halffilled with water. The finer particles are stirred up, and are carried away in the current of water, through the lateral tube, into the beaker placed to receive it; the coarse sand remains in the elutriating glass. The flow of the water is arrested when it runs away nearly clear, and the liquid still in the glass is decanted into the beaker. The coarse sand is washed out into a porcelain crucible, dried, ignited, and weighed.

In about six hours the liquid in the beaker will be nearly clear ; the whole of the fine sand will certainly have been deposited by that time. The supernatant liquid is poured away : the deposit is rinsed back into the narrow glass, and the process of elutriation is repeated. The flow of the water is so regulated that its level in the funnel-tube is about 3 centimetres higher than that in the glass ; in about four hours the whole of the clay proper will have been carried away through the discharge-pipe. The residual fine sand is then rinsed into a weighed porcelain crucible, dried, ignited, and weighed.

The water in the air-dried sample is determined by igniting a second portion for some time ; the amount of the clay proper is determined by difference.

B. *Chemical Analysis.*

The air-dried clay is powdered as finely as possible, a portion is weighed out into a large porcelain crucible, and heated in the steam-chamber for several days. The moisture is calculated from the loss.

About 2 grams of the dried powder are then heated in a platinum dish with excess of moderately-concentrated sulphuric acid for about 8 or 10 hours, and the mass is evaporated to dryness to expel the acid. When cold, the residue is boiled with water, the solution is filtered, and the insoluble matter, consisting of the silica originally existing in union with the bases in the clay, and also of the small quantity of free silica, together with the sand, is washed, dried, and weighed. It is then repeatedly boiled with solution of sodium carbonate in the platinum dish, the liquid filtered, and the residual sand washed with hot water, next with water slightly acidified with hydrochloric acid, and finally with pure water. It is then dried and weighed. Its weight, subtracted from the total weight of the residue left after treatment with sulphuric acid, gives the amount of true silicic acid.

The quantity of uncombined silicic acid in the clay may be determined by boiling a weighed quantity of the sample dried at 100° with a strong solution of sodium carbonate, filtering the liquid, and evaporating to dryness with excess of hydrochloric acid. The separated silicic acid is washed, dried, and weighed. Titanium dioxide is not an infrequent constituent of clays. It may be detected in the residue after treatment with sulphuric acid: an aliquot portion is heated with hydrofluoric and sulphuric acids, whereby the greater portion of the silica is volatilised : the residue is fused with acid sulphate of potassium. Dissolve in cold water, filter if necessary, and boil. The titanium dioxide is reprecipitated: it is to be washed, dried, and weighed. To the filtrate obtained after treatment with sulphuric acid, and containing the bases, add a slight excess of lead nitrate solution, and allow the turbid liquid to stand for a few hours. Filter, wash the lead sulphate, adding the washings to the filtrate, and pass sulphuretted hydrogen through the liquid to remove the excess of lead, again filter, and evaporate the solution to dryness. The alumina, iron, lime, magnesia, and alkalies are obtained as nitrates. Heat gradually to about 250° until no more fumes of nitric acid are evolved. The residue consists of alumina and ferric oxide, calcium, magnesium, and alkaline nitrates. Moisten the residue with a concentrated solution of ammonium nitrate, and heat on the water-bath until no further evolution of ammonia is perceived. Add hot water, filter, wash the alumina and ferric oxide, dry, ignite, and weigh. Transfer the weighed oxides to a porcelain boat, and heat in a stream of dry hydrogen or coalgas for half an hour. Treat the mixture of alumina and reduced iron with very dilute nitric acid (1 pt. acid to 40 of water), warm, filter, and precipitate the ferric oxide by ammonia, wash, dry, and weigh it. The mixture may also be heated with dilute sulphuric acid, filtered, reduced with a small piece of zinc, and titrated with a weak solution of potassium permanganate. Deduct the weight of ferric oxide from that of the mixed oxides: the difference gives the quantity of alumina.

To the filtrate containing the nitrates add ammonium oxalate and weigh the precipitate as caustic lime. Evaporate the filtrate to dryness, ignite to expel the ammonium salts, add excess of oxalic acid (sufficient to convert all the bases present, considered as potash, into quadroxalates), treat with a small quantity of water, and again evaporate to dryness. Ignite gently: the magnesium oxalate is converted into magnesia, and the alkaline oxalates into carbonates. Treat repeatedly with small quantities of water, filter, dry, and weigh the magnesia.

Add a few drops of hydrochloric acid to the filtrate, evaporate to dryness, ignite gently, and weigh the alkaline chlorides. The potash and soda may be separated by platinum tetrachloride (see p. 84), or their proportion may be determined by a dilute standard silver solution in the manner described on p. 127.

The magnesia may also be precipitated by sodium phosphate, and the alkalies determined in a separate portion by ignition with calcium carbonate and ammonium chloride (see Glass, p. 99).

ASSAY OF MANGANESE ORES (PYROLUSITE, BRAUNITE, &c.).

Of the many methods which have been proposed for the valuation of these substances, those of Bunsen and of Fresenius and Will are on the whole the most accurate and convenient. As the value of the oxide depends upon the amount of chlorine which it yields on heating with hydrochloric acid, the method of Bunsen is perhaps the more generally applicable, since it directly determines the amount of chlorine thus evolved. It has the advantage too of being rapidly carried out, and is therefore well adapted to the requirements of manufacturing establishments. When manganese dioxide is brought into contact with hydrochloric acid, the following reaction occurs :

$$MnO_2 + 4HCl = MnCl_2 + Cl_2 + 2H_2O.$$

70·92 parts of chlorine or 253·7 parts of iodine are equivalent to 86·04 parts of manganese dioxide (Mn = 54·04, Schneider). If the chlorine be led into a solution of potassium iodide, an equivalent quantity of iodine is liberated ; this may be determined by the method given on p. 164.

Weigh out 0·5 gram of the finely-powdered ore into the little bulb *a* (fig. 50), add hydrochloric acid, and drive over the chlorine into the U-tube, in which you have previously placed 25 cubic centimetres of the strong solution of potassium iodide.

The remainder of the operation is conducted as described on p. 163.

V.   *Gravimetrical Method of Fresenius and Will.*

This process depends upon the action of manganese dioxide on oxalic acid, in presence of sulphuric acid ; when these substances are brought together, the manganese dioxide parts with an atom of its oxygen to the oxalic acid, which is thereby completely converted into carbon dioxide and water, and the manganese protoxide combines with sulphuric acid to form manganous sulphate ; thus :—

$$MnO_2 + C_2O_2(OH)_2 + H_2SO_4 = MnSO_4 + 2CO_2 + 2H_2O.$$

88 parts of carbon dioxide evolved, correspond therefore to 86·04 parts of manganese dioxide.

The decomposition may be effected in the apparatus represented in fig. 31. From 2 to 4 grams of the finely-powdered ore, according to its supposed richness, are weighed out into the flask A, and covered with dilute sulphuric acid. A strong solution of oxalic acid is then allowed to flow from the bulb-tube, and the evolved carbonic acid collected in the weighed soda-lime tube *c*. It is advisable to place a weighed potash-apparatus (Geissler's form is the most convenient) before the U-tube.

Certain ores of manganese contain considerable quantities of earthy carbonates ; on treatment with an acid their carbonic acid is of course liberated, and thus tends to increase the apparent value of the sample. By means of the apparatus above mentioned, this amount of carbonic acid may be readily determined ; it is only necessary to re-weigh the absorption tubes before the addition of the oxalic acid. Of course, the precaution must be taken to aspirate air through the apparatus before disconnecting the several parts.

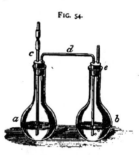

Fig. 54.

The original apparatus devised by Fresenius and Will is represented in fig. 54. It consists of two flasks, *a* and *b*, connected together by a thin glass tube *d*, one end of which terminates just below the cork of *a*, the other ends a few millimetres above the bottom of *b*. The thin tube *c* passes down to the bottom of *a* ; the short tube *e* ends just below the cork of *b*. The two flasks are of about 100 cubic centimetres capacity ; they are made of thin glass, so as to weigh as little as possible. Weigh out into *a* from 3 to 5 grams of the powdered sample, add from 5 to 6 grams of sodium oxalate, and half fill *b* with strong sulphuric acid. Weigh the entire apparatus. Place a short piece of caoutchouc tubing over *c*, and close it by a glass rod, slip a longer piece of caoutchouc tubing over *e*, and aspirate two or three bubbles of air from *a* through the sulphuric acid. On discontinuing the suction, the acid rises in the tube ; if the column remains stationary, the apparatus is air-tight. Now aspirate more air from *a*,

so as to cause a small quantity of sulphuric acid to pass
over from *b.*   Carbonic acid is immediately evolved, and is
dried by passing through the sulphuric acid in *b.*   When the
evolution of gas begins to slacken, draw over in the same
manner fresh quantities of the acid.   The complete decomposition of the ore requires from five to ten minutes ; the termination of the reaction is indicated by the appearance of
the residue in *a,* which will no longer be black, and also from
the non-evolution of carbonic acid when a fresh quantity of
sulphuric acid flows over into *a.*   Remove the glass stopper
from *c,* and aspirate a slow current of air, dried by sulphuric
acid, through the apparatus for five minutes.   When the
flask is cold, remove the caoutchouc tubing, re-weigh, again
aspirate dry air through it, and again weigh.   The two
weighings ought to agree.   The loss of weight indicates the
amount of carbon dioxide evolved by the action of the manganese dioxide upon the oxalic acid.

In the case of ores containing carbonates, treat the
weighed portion in *a* with a little water, add two or three
drops of dilute sulphuric acid, and heat on the water-bath.
In about ten or fifteen minutes test the liquid with a slip of
blue litmus paper ; if it is not acid, add a few more drops
of dilute sulphuric acid, and continue the heating.   When
the carbonates are completely decomposed, neutralise with
caustic soda, free from carbonic acid, allow the liquid to
cool, add the sodium oxalate, connect the flasks together,
and proceed as above.

With proper care this method gives very concordant
results ; two analyses should not differ by
more than 0·25 per cent.

FIG. 54A.

The great weight of the apparatus, and the
surface it exposes, are its chief disadvantages.
To obviate these sources of error, various
modified forms of it have been devised ; one
of the most convenient of these is seen in
fig. 54A.

VI. *Volumetric Determination by means of Iron and Potassium Permanganate Solution.*

About 1·5 to 2 grams of iron wire, perfectly free from rust, are accurately weighed out, and dissolved in 100 c.c. of dilute sulphuric acid (1 of acid to 4 of water), in the apparatus represented in fig. 48, p. 149. About the same weight of the finely-powdered ore to be tested is added to the solution of iron, and the liquid is heated gently, until the whole of the manganese is dissolved. The solution is boiled, the water allowed to recede (see p. 148), and the contents of the flask made up to 250 c.c. When cold the amount of residual ferrous sulphate is determined by permanganate solution :

$$2FeSO_4 + MnO_2 + 2H_2SO_4 = Fe_2(SO_4)_3 + MnSO_4 + 2H_2O.$$

112 parts of iron correspond to 86·04 of manganese dioxide.

Example. — Weighed out 1·562 grams of iron wire. 1·562 × ·996 = 1·556 pure iron. Dissolved in the requisite quantity of dilute sulphuric acid, and added 1·285 gram of manganese dioxide. When cold, the solution required 20·2 c.c. of permanganate solution; 1 c.c. permanganate = 0·01096 Fe. Accordingly, 20·2 × 0·01096 = 0·2215 Fe. 1·556−0·2215 = 1·3345 gram of iron has been oxidised by the manganese dioxide.

$$112 : 86·04 :: 1·3345 : x. \quad x = 1·025$$
and $$1·285 : 1·025 :: 100 : 79·80.$$

The ore, therefore, contained 79·80 per cent. of manganese dioxide.

VII. *Volumetric Determination by means of Oxalic Acid and Potassium Permanganate.*

Since the amount of oxalic acid in solution can be readily determined by means of potassium permanganate (see p. 151), the reaction between the manganic oxide, oxalic acid, and

sulphuric acid may be made the basis of a volumetric
method.   It is merely necessary to heat the finely-powdered
ore with an excess of oxalic acid, and on the completion of the
decomposition to determine the amount remaining in the
liquid by means of standard potassium permanganate solution.
About 2 grams of the ore are weighed out into a flask, and
gently heated with 50 cubic centimetres of normal oxalic acid
solution, and 5 or 6 cubic centimetres of strong sulphuric acid :
when the decomposition is at an end (which may be known
by the absence of any black grains in the sediment), the
solution is filtered into a ½-litre flask, the sediment and filter
washed, and the liquid diluted to 250 cubic centimetres.
After shaking, 100 cubic centimetres of the liquid are introduced into a beaker, and the solution titrated by potassium
permanganate (see p. 151).   The determination is repeated
with a second portion of 100 cubic centimetres.   The mean of
the two results multiplied by 2·5 gives the amount of residual
oxalic acid, and this, subtracted from the amount originally
taken, shows the quantity of acid decomposed by the manganese.   From the equation

$$MnO_2 + C_2O_2(OH)_2 = 2CO_2 + MnO + H_2O$$

it is seen that 90 parts of oxalic acid correspond to 86·04 of
manganese dioxide.

Many samples of manganese ore contain more or less
ferrous oxide, which becomes oxidised at the expense of a
portion of the chlorine evolved on treating the mixture with
hydrochloric acid.   The method of Fresenius and Will, and
the volumetric modification above described, are therefore apt
to assign too high a value to the manganese, the quality of
which, as we have already remarked, depends solely upon
the quantity of available chlorine which it can liberate.

VIII.   *Determination of Moisture in Manganese Ores.*

All manganese ores contain variable amounts of moisture,
the exact determination of which is a point of some

importance. By repeated experiments it has been found that a temperature of 120° maintained for about an hour and a half is sufficient to expel the whole of the hygroscopic moisture, without eliminating any of the true water of hydration. A few grams of the powdered oxide are introduced into the weighed tube, fig. 29, and heated to 120° for about an hour and a half. The loss of weight gives the amount of moisture. If it is preferred to dry the sample at 100° the heat must be maintained for at least six hours before a constant weight will be obtained.

The dried and finely-powdered oxide is exceedingly hygroscopic, and if the sample is dried before being analysed there is great risk that in the operations of weighing the portion will take up fresh quantities of moisture. It is better to keep the powdered sample undried, in a well-corked test-tube, and to weigh out portions for the determination of the oxide and water at the same time, and to conduct the two operations simultaneously. The percentage amount of oxide in the sample when dried may be afterwards readily calculated.

IX. *Determination of the Amount of Hydrochloric Acid required to decompose Manganese Ore.*

Two samples of manganese ore may show on analysis the same amount of available dioxide—that is, may liberate the same amount of free chlorine and yet require very different amounts of hydrochloric acid to effect their complete decomposition. The elimination of a given quantity of chlorine by means of hausmannite, for example, requires the expenditure of twice the amount of hydrochloric acid needed to yield the same quantity from the binoxide :

$$Mn_3O_4 + 8HCl = 3MnCl_2 + Cl_2 + 4H_2O.$$
$$MnO_2 + 4HCl = MnCl_2 + Cl_2 + 2H_2O.$$

Moreover, since most ores of manganese contain carbonates, and gangue, decomposable by hydrochloric acid, it is always necessary to employ a larger quantity of acid than corresponds to the amount of available binoxide found. Accordingly it is often required to determine exactly the quantity of acid needed to effect the complete decomposition of the ore.

Dissolve a quantity of recrystallised copper sulphate in warm water, and carefully add solution of ammonia, with constant stirring, until the bluish-green precipitate is very nearly dissolved. If the exact point is overstepped, add a little more solution of copper sulphate until the precipitate just reappears. Filter the solution, and determine its value by withdrawing 10 cubic centimetres, and adding standard sulphuric acid until the liquid remains permanently turbid. Now weigh off about 1 gram of the powdered manganese ore into a small flask fitted with a cork, in which is fixed an obtusely-bent tube about 3 feet in length, and add to it 10 cubic centimetres of moderately-concentrated hydrochloric acid (sp. gr. 1·1), and heat gently ; the bent tube is to be so arranged that the condensed water flows back into the flask. Whilst the liquid is heating, measure off a second quantity of 10 cubic centimetres of the acid, run it into a beaker, and add the standardised ammonio-copper solution, with constant stirring, until the liquid is just rendered turbid. Heat the contents of the flask more strongly for a few minutes to expel the chlorine; let the flask cool, add a quantity of cold water to it, throw the solution on to a filter, wash, and again titrate with the copper solution. The difference expresses the amount of hydrochloric acid required to decompose the ore. As the amount of chlorine evolved is known from a determination made by one of the preceding methods, the quantity remaining in combination as manganous chloride, &c., is readily calculated. This method, although not absolutely exact, affords results of sufficient accuracy for technical purposes. The ordinary acidimetric methods are here inapplicable, since the manganous chloride possesses an acid reaction which interferes with the process. This solution (known as Kieffer's) is frequently of service in testing liquors containing free acid or salts which redden litmus ; in determining, for example, the amount of free acid in liquids from galvanic batteries, &c. It may be also used in determining the strength of vinegars, as the brown colour of the solution in no way interferes with the completion of the reaction. It is necessary from time to time to redetermine the strength of the solution by standard acid, as it experiences slight alteration on keeping.

X. BLEACHING POWDER.

Bleaching powder or chloride of lime is formed by the action of chlorine upon calcium hydrate. The relation of its constituents in the freshly-prepared substance is represented by the formula $Ca_3H_6O_6.Cl_4$. When allowed to stand in contact with air and light, chloride of lime suffers decomposition, and, after treatment with water, the calcium chloride is found to have increased in quantity, whilst the hypochlorite has suffered a corresponding diminution. When exposed to moist air containing carbonic acid, bleaching powder is decomposed, hypochlorous acid is evolved, and calcium carbonate formed. When, therefore, chloride of lime is used as a disinfectant, the active agent in ordinary circumstances is hypochlorous acid, and not free chlorine, as formerly supposed. At a moderate temperature ($50°$) dry chloride of lime is converted into calcium chlorate, and the mass becomes pasty from the separation of water :

$$3Ca_3H_6O_6Cl_4 = 5CaCl_2 + Ca(ClO_3)_2 + 3CaH_2O_2 + 6H_2O.$$

This change goes on at a diminished rate even in direct sunlight. Chloride of lime is decomposed by water, calcium hydrate separates out, and calcium chloride and hypochlorite pass into solution :

$$Ca_3H_6O_6Cl_4 = CaH_2O_2 + CaCl_2 + Ca(ClO)_2 + 2H_2O.$$

It is highly probable that the hypochlorite thus formed

is only produced by the action of the water, and does not exist pre-formed in the bleaching powder.

Since the value of the commercial article depends entirely upon the amount of hypochlorous acid which it can produce, and since the circumstances of heat, moisture, air, and light exercise such an important influence upon the proper production and stability of the bleaching powder, it is evident that, as manufactured and stored, it must vary very considerably in quality. The most concentrated preparation which can be obtained by saturating calcium hydrate with chlorine, contains about 38·5 per cent. of available chlorine, but the great bulk of the substance found in commerce rarely contains more than from 32 to 37 per cent., of which 1 or 2 per cent. is without bleaching power, being present in the form of calcium chlorate. In badly-made bleaching powder the amount of chlorate present is occasionally equal to 8 or 10 per cent. of available chlorine—nearly one-fourth of the amount which ought to be contained in the product. Many methods have been proposed to estimate the available chlorine present in bleaching powder, the majority being based on the oxidising effect of the hypochlorites, but a great number are inaccurate, in that they do not take cognisance of the presence of this admixed chlorate, which, under the circumstances of the valuation-processes, reacts like chlorine, although it has no bleaching effect.

The best and most convenient chlorimetrical methods hitherto proposed are those of Penot and Bunsen.

Penot's Method.—This process is based upon the conversion of an alkaline arsenite, by the chloride of lime solution, into an arseniate :

$$As_2O_3 + Ca(ClO)_2 = As_2O_5 + CaCl_2.$$

The final point of the reaction is determined by means of potassium iodide and starch ; so long as any hypochlorite remains undecomposed, a drop of the solution brought into contact with potassium iodide and starch renders that mixture blue. This mixture of iodide and starch is conveniently employed in the form of test-papers. 3 grams of arrowroot, potato, or wheat starch are rubbed into a thin cream with 50 or 60 cubic centimetres of warm water. Pour the mixture into about 200 cubic centimetres of water, and heat the liquid, with constant stirring, until it boils: now add 1 gram of potassium iodide and 1 gram of pure carbonate of soda dissolved in a little water, and dilute the mixture to 500 cubic centimetres. Moisten a number of strips of Swedish filter-paper, or other unsized paper of good quality, with the solution, and when dry, preserve them in a wide-mouthed stoppered bottle. To prepare the arsenious acid solution, powder a quantity of the purest sublimed arsenious acid (free from arsenic sulphide), and weigh off exactly 4·95 grams into a litre flask, add about 25 grams of recrystallised sodium carbonate (free from sodium sulphide, sulphite, or thiosulphate) and 200 cubic centimetres of water. Boil the solution gently, and shake it continually until the arsenious acid is dissolved: when the solution is cold dilute it exactly to one litre. This constitutes a deci-normal solution of arsenious acid: the equivalent of As_2O_3 is 198. 1 eq. can take up 2 atoms of oxygen to form As_2O_5, or is equivalent to 4 of Cl. Since it is difficult to weigh out exactly the required quantity of arsenious acid, it is preferable to take a round number, about 5 grams, and dilute proportionally.

Example.—5·013 grams were weighed out into the litre flask, 25 grams of sodium carbonate and 200 cubic centimetres of water added: after complete solution and cooling the liquid was diluted to 1 litre, and 12·7 cubic centimetres of water were added by means of a burette ; since

$$4·95 \; : \; 1000 \; :: \; 5·013 \; : \; 1012·7.$$

The solution in the flask is well shaken, and decanted off into a number of small well-stoppered bottles : this precaution diminishes the liability of the solution to change on exposure to the air. If the solution is perfectly free from sodium thiosulphate, or sulphite, sodium or arsenic sulphides, there is far less chance of it suffering alteration.*

The sample of bleaching powder to be tested is well mixed, and about 10 grams are weighed out into a porcelain mortar ; 50 or 60 cubic centimetres of water are added, and the mixture is rubbed to a thin cream ; it is allowed to settle for a few minutes, and the supernatant liquid (which is still turbid) poured into a litre flask. The sediment in the mortar is triturated with fresh water, and the operation is repeated until the whole of the chloride of lime has been brought into the litre flask. Fill up to the mark, and shake. Have the burette ready filled to the zero mark, withdraw 50 cubic centimetres of the turbid solution, run it into a beaker, and add the arsenious acid solution, with constant stirring, until a drop from the beaker, taken out on a glass rod, and brought into contact with a strip of the iodised paper moistened with water on a white plate, no longer gives a blue stain. There is no difficulty in hitting the final point ; the gradually increasing faintness in the blue colour of the drops indicates with great accuracy the progress of the reaction. In making a second determination, care must be taken to shake the contents of the litre flask before withdrawing the solution ; if this precaution be neglected, the second determination will give a much lower result—a difference of 2 or 3 cubic centimetres being not unfrequently obtained in testing the clear and the turbid liquids.

Example.—10·99 grams of bleaching powder were treated as directed, and diluted to 1 litre. 50 cubic centimetres of

* When a great number of chlorimetrical estimations have to be made it will be found convenient to modify the above method in the following manner :—The weighed quantity of arsenious acid is dissolved by a gentle heat in 10 or 15 c.c. of *glycerine*, and diluted with water to 1 litre. The weighed sample of bleaching powder is treated with water as directed, and a portion of the *turbid* solution poured into a burette. 25 c.c. of the standard arsenious acid solution are delivered into a flask, mixed with 1 cubic centimetre of *indigo solution*, and the bleaching powder solution added, with constant shaking, until the blue colour is discharged.

the turbid solution required 47·3 cubic centimetres of the arsenious acid solution to complete the reaction. Since 1 cubic centimetre of this solution is equivalent to 0·003546 of chlorine, this would correspond to 47·3 × 0·003546 = 0·1677 gram in the 50 cubic centimetres of solution. But the 50 cubic centimetres contain 0·5495 gram of the bleaching powder ; hence the substance contains $\dfrac{0·1677 \times 100}{0·5495} =$ 30·52 per cent. of chlorine.

The amount of calcium chloride present in a sample of bleaching powder may be determined by first estimating the hypochlorite in the manner above described, and then adding to a second portion of 50 cubic centimetres a slight excess of ammonia, and warming. The hypochlorite is thus converted into the chloride, with the formation of water and nitrogen :

$$3Ca(ClO)_2 + 4NH_3 = 3CaCl_2 + 6H_2O + 4N.$$

The chlorine is then determined in the solution, after boiling, and cautiously neutralising with nitric acid, by means of standard silver solution. In normal bleaching powder the amount of available chlorine will be equivalent to that existing as calcium chloride. To determine the amount of chlorate present, a third portion is heated with ammonia, strongly acidified with pure sulphuric acid, and digested with metallic zinc. In a few hours the nascent hydrogen will have completely reduced the chloric acid to the state of hydrochloric acid, and on again precipitating the chlorine, the increased amount over the second determination shows the quantity existing as chlorate.

Another method of estimating the total chlorine present in bleaching powder, is to boil the turbid solution with a solution of ferrous sulphate and potash, whereby the hypochlorous and chloric acids are reduced to hydrochloric acid :

$$HClO + HClO_3 + 8FeO = 2HCl + 4Fe_2O_3.$$

The solution is filtered, acidified with nitric acid, and the chlorine precipitated with silver nitrate.

In some parts of the Continent, particularly in France, it is customary to represent the amount of available chlorine, not in percentages, but in chlorimetrical degrees, representing the number of litres of chlorine at 0°, and 760 millimetres which 1 kilo. of the sample should yield. Thus, if a sample is reported to be of 100°, it means that 1 kilo. of it would yield 100 litres of chlorine, measured at the standard temperature and pressure. Since a litre of chlorine weighs 3·177 grams, this sample would contain in a kilo. 317·7 grams, or 31·77 per cent. Conversely it is easy to see that 31·77 per cent. would be equal to 100°, since 31·77 per cent. is equal to 317·7 per mille, and $\dfrac{317\cdot7}{3\cdot177} = 100°$.

Bunsen's Method (Modified.)—Withdraw 20 cubic centimetres of the turbid solution, made in the manner above described, and place it in a beaker, add about 15 cubic centimetres of potassium iodide solution, acidify with hydrochloric acid, and determine the iodine liberated according to the method given on p. 157. The amount of available chlorine in the sample is thus measured by the quantity of iodine which it can set free.

XI. BLACK-ASH ; SODA-ASH ; VAT-WASTE.

Black-ash is the product obtained by heating the mixture of sodium sulphate (salt-cake), calcium carbonate, and small coal or slack in a reverberatory furnace, in the manufacture of soda by Leblanc's process. It consists essentially of a mixture of carbonate and caustic soda with sulphide and carbonate of calcium. In addition it contains small quantities of sodium sulphite, thiosulphate (hyposulphite), sulphide, and undecomposed sulphate and chloride, together with alumina, ferrous sulphide, sand, and unburnt carbon; to the last-named substance is mainly due the characteristic colour of the product.

On lixiviating the fused mass with tepid water, the greater

portion of the sodium compounds pass into solution, and on evaporating the clear liquid crude *soda-ash* is obtained. The insoluble matter remaining in the lixiviating tanks, and consisting mainly of calcium sulphide and carbonate, is termed *tank-* or *vat-waste.*

The difference in composition of these various products is well seen in the following analyses :—

I. BLACK-ASH—GERMAN MAKE (ANALYSED BY FRESENIUS).

	Sodium carbonate	.	31·982
	Sodium hydrate	.	6·104
	Sodium silicate .	.	1·019
Soluble in water	Sodium aluminate	.	1·080
	Sodium sulphide	.	0·133
	Sodium sulphite	.	0·216
	Sodium chloride	.	0·288
			40·822

	Calcium sulphide *	.	32·342
	Calcium carbonate	.	10·234
	Lime	.	8·900
	Ferrous sulphide	·.	0·916
Insoluble in water	Silica	.	0·377
	Alumina	.	0·671
	Soda	.	0·641
	Carbon	.	3·528
	Sand	.	1·417
			59·026
			99·848

II. REFINED SODA-ASH—GLASGOW (ANALYSED BY BROWN).

Sodium carbonate	.	80·92
Sodium hydrate	.	3·92
Sodium silicate .	.	1·32
Sodium aluminate	.	1·01
Sodium sulphate	.	7·43
Sodium sulphite	.	1·11
Sodium thiosulphate	.	trace
Sodium sulphide	.	0·23
Sodium chloride	.	3·14
Insoluble matter	.	0·77
		99·85

* The combination of lime and sulphur in the portion insoluble in water has been rearranged in accordance with the views now generally held as to the manner in which these bodies are united in the vat-waste. In the original analyses the sulphur (excluding that present in the ferrous sulphide) was calculated to calcium oxysulphide, $3CaS.CaO$.

The following scheme gives the method for the complete analysis of black-ash.

About 30 grams of the finely-powdered ash are digested with water at a temperature of about 45° in a flask of 500 c.c. capacity. The solution should be hastened as far as possible by repeated shaking ; the insoluble matter is then allowed to subside, and in a few hours the clear supernatant liquid is poured through a folded filter into a ½-litre flask. The residue, which should be kept as far as practicable in the flask in which it was originally placed, is quickly washed with cold water, and the washings added to the main bulk of the filtrate. Discontinue the washing when the filtrate commences to be turbid, and fill up the ½-litre flask to the containing-mark. Close and shake it. If the knowledge of the nature of the soluble matter is not immediately wanted, it is better to proceed at once with the examination of the insoluble portion, since this is apt to suffer alteration on standing.

A.   *Analysis of Insoluble Matter.*

Wash the precipitate from the filter back again into the flask in which the main portion of the residue is contained, and without delay attach the flask to the rest of the apparatus represented in fig. 55, the several parts of which should be already weighed, as indicated below, and put together. A is the flask containing the insoluble matter. It is fitted with a caoutchouc cork, through which is passed a 100 c.c. pipette filled with hydrochloric acid of sp. gr. 1·1. To the upper end of the pipette is attached a piece of caoutchouc tubing which can be closed by a screw clamp. The other end of the tube is connected with a soda-lime tube, S ; A is connected with the flask B, of 300 c.c. capacity, and containing a cold saturated solution of copper acetate free from sulphuric acid. Both flasks stand on an iron plate which can be heated by a lamp; B is connected with the two-bulbed U-tube c, also containing copper acetate solution, standing in an empty

beaker, and fitted with bent tubes in such a manner that
the gas traversing the apparatus passes twice through the

FIG. 55.

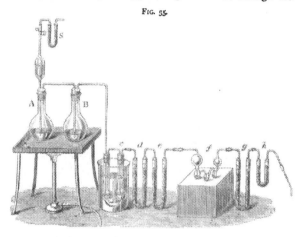

same quantity of liquid (fig. 56). The two U-tubes *d* and
e are filled with calcium chloride to dry the gas : *f* is a
potash-apparatus (of the form known as
Geissler's); it is filled to the extent indicated in the figure with solution of potash
of sp. gr. 1·27 (containing about 30 per
cent. of potassium hydrate), and is accurately weighed : *g* and *h* are U-tubes
filled with soda-lime and calcium chloride :
g only is weighed, its object is to absorb
the last traces of carbon dioxide; *h* is
placed merely to prevent the absorption
of atmospheric carbon dioxide and moisture by *g* : the
caoutchouc tube at the end leads to the filter-pump or
other aspirating arrangement. As soon as the apparatus is
put together, fill the pipette with hydrochloric acid, fit the .

FIG. 56.

cork into A and open the clamp, so as to cause the acid to enter the flask. The insoluble matter in A is immediately decomposed, and sulphuretted hydrogen and carbon dioxide are simultaneously evolved; the former is absorbed in B and c; the latter, after being dried by passing through d and e, is absorbed by f and g. When the evolution of the gas slackens, add more acid, until the decomposition is complete. Heat the iron plate until the contents of A and B are in gentle ebullition, pour hot water into the beaker, open the clamp, and aspirate a slow current of air (about 5 litres) through the apparatus.

The increase in the weight of f and g gives the amount of carbon dioxide. The copper sulphide is thrown on to a filter, and, without washing, the precipitate, together with the filter, is transferred to a flask and treated with hydrochloric acid and potassium chlorate. The liquid is filtered into a litre flask, the filter washed, and the filtrate diluted to the containing-mark. Withdraw two portions of 100 c.c., and determine in each the sulphuric acid by means of barium chloride. 233·2 parts of barium sulphate are equivalent to 32 parts of sulphur.

Pour the contents of the flask A on to a weighed filter, receiving the filtrate in a $\frac{1}{2}$-litre flask, wash the insoluble matter, consisting of sand and carbon, dry at 100° and weigh. Ignite the weighed mixture to burn off the carbon, and weigh the residual sand. The difference between the two weighings gives the carbon.

Make up the filtrate to 500 c.c., and transfer 200 c.c. to a porcelain basin, add a small quantity of nitric acid, and evaporate to dryness on the water-bath. Treat the separated silica in the usual manner and weigh it. Precipitate the iron and alumina from the filtrate by ammonia, weigh them together, fuse the mixture with a little acid potassium sulphate, and determine the iron volumetrically by potassium permanganate. The alumina is found by difference. Determine the lime, magnesia, and alkalies by the method given on p. 184.

B. *Analysis of the Soluble Portion.*

1. Withdraw 50 c.c., and determine the total alkali present by standard acid, litmus, and soda, in the manner directed on p. 139. Calculate as sodium carbonate.

2. Transfer 100 c.c. to a ½-litre flask, mix with barium chloride solution so long as a precipitate forms, fill up the flask to the containing-mark, shake, and close it. Determine the sodium hydrate in an aliquot portion of the clear liquid by standard acid and litmus.

3. Transfer 50 c.c. to a large beaker, dilute with 200 c.c. of water, add acetic acid until the liquid is very nearly neutral, and determine the joint amount of the sulphide and sulphite of sodium by means of starch paste and standard iodine solution.

4. Transfer 100 c.c. to a ½-litre flask, and add zinc sulphate solution made strongly alkaline by potash, until a considerable precipitate is formed. This contains the whole of the sulphur present as sodium sulphide ; the rest of the sulphur remains in the solution. Dilute to the containing-mark, shake, allow to settle, and determine the sulphur still in solution in an aliquot portion of the clear liquid by means of starch and standard iodine solution, after acidifying with acetic acid in the manner previously directed. From the amount of iodine used, the sulphur dioxide is readily calculated : 253·7 parts of iodine are equivalent to 126·15 of sodium sulphite. The difference between this and the previous determination of the total amount of iodine required gives the quantity of sodium sulphide : 253·7 parts of iodine correspond to 78·15 of sodium monosulphide.

5. Evaporate 100 c.c. to dryness in a thin porcelain basin with a small quantity of pure nitre, and gently fuse the residue. The sodium sulphide and sulphite are thereby oxidised to sulphate. Digest the mass with hot water, filter the solution into a 250 c.c. flask, wash the insoluble matter, and fill up the flask to the mark, and shake. Withdraw 100

c.c., and determine the sulphuric acid, after acidulating with hydrochloric acid, as barium sulphate. Subtract the amount corresponding to the sodium sulphide and sulphite ; the remainder is calculated to sodium sulphate.* In another 100 c.c. determine the chlorine by standard silver.

6. Transfer 100 c.c. to a porcelain dish, acidify with hydrochloric acid, and separate the silica in the usual way. In the filtrate determine the alumina by precipitation with ammonia.

Arrangement of the Results.—A. *Insoluble Portion.*—Calculate the iron to ferrous sulphide, and the remainder of the sulphur to calcium sulphide, CaS. The rest of the calcium is to be set down as lime, CaO. The silica, alumina, and soda are set down uncombined, as we have no knowledge of their state in the insoluble portion. The sand and carbon are, of course, directly determined.

B. *Soluble Portion.*—Combine the silica and alumina with soda, to form sodium silicate and aluminate, Na_2SiO_3, and $Na_2Al_2O_4$; calculate the amount of sodium carbonate equivalent to these compounds, together with that corresponding to the hydrate and sulphide, and subtract the joint amount from the result obtained by determining the total alkali with standard acid and soda. The remainder gives the real sodium carbonate present in the solution.

The foregoing scheme of analysis is, of course, equally applicable to the analysis of soda-ash and vat-waste. Vat-

---

* It is sometimes requisite for the purposes of the manufacturer merely to determine the quantity of sodium sulphide in the liquor. This may be readily effected by the following process, due to Lestelle. The solution is mixed with ammonia, heated to boiling, and titrated with a weak standard solution of silver, made by dissolving 4·3475 grams of pure silver nitrate in water in a litre flask, adding excess of ammonia, and diluting to 1,000 c.c. 1 c.c. is equivalent to 1 milligram of sodium sulphide. When the sulphur is nearly all precipitated, the liquid is filtered, and the addition of the silver solution continued until, after again filtering, only the faintest turbidity is produced. The method is expeditious, and after a little practice gives accurate results.

waste is treated exactly like the insoluble portion of black-ash ; soda-ash like the soluble portion.

XII. ESTIMATION OF SULPHUR IN PYRITES, BY MEANS OF COPPER OXIDE.

(Particularly applicable to roasted Pyrites.)

A rapid and sufficiently accurate method for technical purposes consists in heating from 5 to 10 grams of the ore, according to its richness in sulphur (if more than 10 per cent. of sulphur is present, 5 grams, if less than 10 per cent., 10 grams are taken), intimately mixed with 5 grams of pure sodium carbonate, and about 50 grams of dry copper oxide, in a porcelain crucible, to a low red heat, for about a quarter of an hour, with frequent stirring. The sulphur is oxidised to sulphuric acid, and combines with the alkali. When cold the mass is treated with water, and the amount of sodium carbonate remaining determined by titration with normal acid.

XIII. ASSAY OF COPPER ORES (MANSFELD PROCESS).

About 5 grms. of the finely-powdered ore are weighed out into a flask, and mixed with 40 c.c. of moderately-concentrated hydrochloric acid (sp. gr. 1·16). 6 c.c. of dilute nitric acid (made by mixing equal bulks of water and pure acid of sp. gr. 1·2) are added, and the flask is gently heated for 30 minutes on a sand-bath, after which it is boiled for 15 minutes. The whole of the copper is now in solution : the extraction is complete, even in the case of very rich ores, provided sufficient attention has been paid to the powdering. The solution is filtered into a large beaker, into which a rod of zinc, weighing about 50 grms., and surrounded with a piece of thick platinum foil, has been previously placed. It is necessary that the zinc employed should be as free as possible from lead. The precipitation of the metallic copper commences immediately, and is generally complete in about half an hour. The rod of zinc is withdrawn, and the precipitated copper repeatedly washed by decantation. If the amount of the copper does not exceed 6 per cent. (which may be approximately known from the bulk of the reduced metal), it is dissolved in 8 c.c. of the dilute nitric acid, prepared as above. The beaker is gently warmed, and the amount of copper in the liquid titrated by solution of potassium cyanide, after previous addition of 10 c.c. of ammonia solution, prepared by diluting 1 vol. of ammonia-water (sp. gr. 0·93) with 2 vols. of water. When the amount of copper in the ore exceeds 6 per cent., the metal is dissolved in 16 c.c. of the nitric acid solution, and the liquid is washed into a 100 c.c. flask, diluted to the containing-mark, shaken, 50 c.c. withdrawn, mixed with 10 c.c. of the dilute ammonia, and titrated with potassium cyanide. The experiment may be repeated with the second portion of 50 c.c.

When a solution of potassium cyanide is mixed with an ammoniacal solution of copper sulphate or nitrate, the azure blue colour gradually disappears with the formation of copper-ammonium-cyanide, free ammonium cyanide, ammonium formate, and urea. The reaction is only constant so long as the amount of free and combined ammonia present is invariable.

The strength of the solution of the potassium cyanide is thus tested :—Exactly 5 grams of chemically pure copper, prepared by the electrotype process, are weighed out into a litre flask, and dissolved at a gentle heat in 266·6 c.c. of the dilute nitric acid (made in the manner above described). On cooling, the solution is diluted to the containing-mark. 30 c.c. of this solution, containing 0·15 grm. of metallic copper, are placed in a beaker and mixed with 10 c.c. of the dilute ammonia liquid, and the solution of potassium cyanide is added from a burette, with constant stirring, until the blue colour of the liquid just disappears. The strength of the cyanide should be so arranged that 1 c.c. of the solution is equivalent to 5 milligrams of copper. The titration

of the solution of the sample of ore is made in exactly the
same manner. If exactly 5 grams have been taken, and the
cyanide is of the above strength, each cubic centimetre of
the solution required for decolourisation is equivalent to o·1
per cent. of copper. The number of cubic centimetres
needed, divided by 10, gives the percentage of copper at
once.

This method is very expeditious, and if due care be exer-
cised, it is very accurate. It must be borne in mind that it
is strictly comparative, and the titrations must be made
therefore in a uniform manner. The presence of a very
small quantity of lead exercises no influence on the results,
but the action of zinc is more injurious. Care must be taken
therefore to wash the precipitated copper thoroughly before
dissolving it in the dilute nitric acid. The solutions must
be quite cold before titration, since less potassium cyanide
is needed to decolourise a solution when warm than when
cold. Thus while 30 c.c. of copper solution, containing o·15
grm. copper, and 10 c.c. normal ammonia solution, required
at the ordinary temperature exactly 30 c.c. of potassium
cyanide solution, the same quantities at about 45° required
only 28·9 c.c. (STEINBECK).

The solution of potassium cyanide requires to be titrated
from time to time, since its strength is not invariable.

XIV. ASSAY OF COPPER ORES (LUCKOW'S PROCESS).*

This process, which is now largely used in many German
establishments, depends upon the fact that copper is precipitated in the metallic state from acid solutions by a weak
galvanic current. The operations required by this method
may be best described under the following heads : 1. Roasting the ore ; 2. Solution of the roasted ore ; 3. Precipitation
of the copper ; 4. Weighing the copper.

1. *Roasting the Ore.*—This operation is only necessary
when the ores are bituminous. Weigh out about 2 grams of

* Zeitsch. für Anal. Chemie, Fresenius, 1869, p. 1.

the finely-powdered sample into a thin porcelain crucible, and heat it over a Bunsen flame for 10 minutes, occasionally stirring it with a thick platinum wire, so as to expose fresh surfaces to the oxidising action of the air. The bituminous matter and the greater portion of the sulphur will be expelled at the expiration of this time.

2. *Solution of the Ore.*—The roasted powder is transferred to a small flat-bottomed beaker, about 5 centimetres in height and 3 centimetres wide, and treated with 6 c.c. of nitric acid of sp. gr. 1·2, 4 c.c. of a dilute sulphuric acid (prepared by mixing equal volumes of the strong acid and water), and 25 drops of hydrochloric acid. The addition of the sulphuric acid increases the oxidising action of the nitric acid, and converts any lime which may be present into the difficultly-soluble calcium sulphate. The liquid is evaporated to complete dryness on a sand-bath, the beaker being meanwhile covered with a funnel, the stem of which has been cut off. This operation requires about an hour. The addition of the hydrochloric acid facilitates the evaporation and decreases the tendency of the liquid to spirt. When dry, break up the mass with a glass rod.

3. *Precipitation of the Copper.*—Wash the cover inside and out with dilute nitric acid (1 vol. of acid of sp. gr. 1·2 diluted with 6 vols. of water), and also the sides of the beaker, until it is about half-filled with the dilute acid. Add a few drops of a strong solution of tartaric acid (which is best preserved in a beaker simply covered with a piece of paper) to the liquid, and place the spiral, represented in fig. 56*a*, within the beaker. This spiral consists of a piece of platinum wire about 18 centimetres long, and 1 millimetre thick ; two-thirds of it are so bent in circles that the straight portion of the wire projects as if it were the axis of the spiral. The outer convolution is so large that it just touches the sides of the beaker : the vertical portion of the wire is therefore exactly in its centre. If the evaporation has been carefully attended to, the acid liquid remains quite clear :

should it be turbid, add 1 c.c. of a concentrated solution of barium nitrate, and agitate the mixture by the aid of the platinum spiral. A piece of stout platinum foil, to which a thick platinum wire has previously been attached, is then bent into a cylinder of 3 centimetres long and 18 millimetres in diameter (fig. 56 *b*); this is accurately weighed, and supported in the beaker about 1 millimetre from the spirals: the vertical portion of the spiral becomes therefore the axis of this

FIG. 56*a*.   FIG. 56*b*.   FIG. 56*c*.

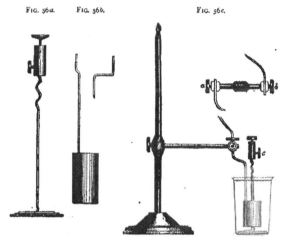

cylinder. The wire supporting the foil is fixed by means of a screw, *a*, to the arm *a b* of the stand (fig. 56 *c*), the other screw, *b*, holds the wire leading from the zinc pole of a constant battery. A small screw clamp, *c*, is fastened to the end of the platinum spiral, and connects the arrangement with the other pole of the battery. Immediately the circuit is closed, copper commences to be deposited on the platinum foil, and bubbles of oxygen are given off from the spiral: in about 8 hours the whole of the metal will certainly be precipitated, even from rich ores and by a weak current. A stream of water is run into the beaker so as to displace the acid liquid, which is allowed to flow over the sides. The platinum cylinder is then withdrawn and disconnected from the stand, washed with hot water, and then with a few drops of alcohol. It is heated in the steam-bath, and weighed when cold. Its increase in weight gives the quantity of metallic copper present in the sample. As the process of precipitation needs no superintendence, it may be allowed to go on during the night: if the operation be commenced in the evening, the reduced copper will be ready for washing and weighing in the morning. The copper should show its characteristic red colour, and be free from any saline deposit : the absence of this deposit is evidence of the perfect removal of copper from the liquid.

XV. Assay of Copper Ores by Precipitating the Metal by means of Zinc.

(See p. 105.)

About 5 grams of the finely-powdered ore are gently heated with strong aqua regia in a deep porcelain crucible, covered with a watch-glass. The glass is rinsed with water, and the liquid is evaporated to complete dryness. The dried residue is heated, to expel the unoxidised sulphur ; if the ore contains much pyrites, it will be necessary to treat it again with strong nitric acid (B.P. 86°), and evaporate a second time to dryness, and roast.

Treat the dried mass with hot water to extract the copper (now present as sulphate), filter into a weighed platinum dish, wash the insoluble residue, adding the washings to the solution already in the dish, and precipitate the copper by means of zinc, in the manner directed on p. 105. To prove the accuracy of this method, 1 gram of pure metallic copper was mixed with 0·5 gram of the following substances, either in the metallic state, or as salts, viz. :—Gold, silver, platinum, tin, lead, iron, zinc, nickel, cobalt, bismuth, arsenic, uranium, mercury, molybdenum, antimony, sulphur, silica, and calcium phosphate. The mixture was treated in the manner described, and the copper precipitated by zinc ; the amount of reduced copper was 0·996 gram. In more than twenty determinations of copper in various combinations, the average amount of the metal obtained by this method was 99·7 per cent. of the actual quantity present. (MOHR.)

XVI. COPPER PYRITES.

This mineral constitutes the most abundant ore of copper : it consists essentially of sulphur, iron, and copper, and when pure contains 34·8 per cent. of copper. It is almost invariably mixed, however, with more or less antimony, arsenic, bismuth, lead, manganese, zinc, nickel, and cobalt in addition to considerable quantities of silicious substances and gangue. When a complete quantitative analysis of the ore is to be made, it must always be preceded by a careful qualitative examination.

Pulverise about 20 grams of the mineral in an agate mortar, and dry at 100°.

Determination of the Sulphur.—About 1 gram of the ore is weighed out into a porcelain dish, a few crystals of potassium chlorate are added, together with about 50 cubic centimetres of pure nitric acid (sp. gr. 1·35), and the mixture is heated on the water-bath. To prevent loss by spirting, the dish should be covered with a large watch-glass or funnel. Fig. 57. As the evolution of chlorine diminishes, add occasionally a crystal of potassium chlorate. In about an hour the whole of the sulphur will be oxidised ; the cover is rinsed with hot water, and the liquid in the dish concentrated to a small bulk, a small quantity of strong hydrochloric acid is added, and the solution is evaporated to perfect dryness, in order to render the silica insoluble. Moisten the dried residue with strong hydrochloric acid, add hot water, filter, and wash the insoluble portion by decantation. This consists mainly of silica and undecomposed gangue; it may contain, however, a small quantity of lead sulphate, left undissolved by the hydrochloric acid. The last traces of the sulphate may be removed by heating the mass with a solution of ammonium acetate (made by mixing solutions of ammonia and acetic acid), and adding the liquid to the main quantity of the filtrate. Add to the solution a few crystals of tartaric acid, to prevent the precipitation of traces of iron, heat the liquid to boiling, and add excess of barium chloride. Allow the liquid to stand; filter, wash the precipitate by decantation with hot water, and digest it with a solution of ammonium acetate, which removes any traces of barium nitrate, which the precipitate is very apt to retain. The barium sulphate is then washed on to the filter, dried, and weighed.

Fig. 57.

Determination of the Copper, Iron, &c.—2 or 3 grams of the powdered ore are oxidised with fuming nitric acid in a dish, a few cubic centimetres of strong sulphuric acid are added, and the liquid evaporated to dryness. The residue is dissolved in hydrochloric acid, water added, the liquid allowed to stand, and then filtered into a flask. The insoluble portion consists mainly of silica and gangue, but it may still retain a small quantity of lead as sulphate; this is to be removed by repeatedly boiling the residue in the dish with dilute hydrochloric acid, and passing the liquid through the filter.

The residue is then transferred to the filter, washed with hot water, dried, and weighed. The filtrate is warmed to about 70°, and a current of sulphuretted hydrogen passed through it. The copper, lead, bismuth, tin, arsenic, and antimony are precipitated. Allow the liquor to stand, pour

the supernatant liquid through a filter, and wash by decantation with water containing a little sulphuretted hydrogen. The small portion of the sulphide adhering to the
filter is washed back into the flask, and the precipitate is
gently heated with a moderately concentrated solution of
potassium sulphide ; after a short digestion (thirty or forty .
minutes) add water (otherwise a little copper may remain
in solution), and again filter. The sulphides of arsenic and
antimony, mixed with sulphur, obtained by adding hydrochloric acid to the filtrate, are filtered off, and oxidised by red
fuming nitric acid (B.P. 86°), the solution concentrated, an
excess of sodium carbonate added, and the whole evaporated to complete dryness, and fused in a silver dish. The
fused mass is further treated as in No. XV. Part II.

The sulphide of copper containing the small quantities of
lead and bismuth, is dried and transferred to a small porcelain basin, and dissolved in nitric acid. The solution is
evaporated to a small bulk, chloride of ammonium added to
dissolve the bismuth, and then dilute sulphuric acid. Allow
the precipitate to settle completely, pour off the clear liquid
through a filter, quickly wash twice or three times with a
little water containing a drop or two of sulphuric acid, rinse
the lead sulphate on to the filter by means of alcohol, and
wash the filter paper thoroughly with alcohol. Do not mix
the alcoholic washings with the main quantity of the filtrate.
The bismuth is best precipitated as carbonate. Nearly
neutralise the filtrate containing the copper, and add excess
of ammonium carbonate, gently warm, filter, dissolve the
precipitate, which consists partly of bismuth carbonate,
partly of basic bismuth sulphate mixed with copper, in a
few drops of nitric acid, and again add ammonium carbonate, which reprecipitates the bismuth as pure carbonate free
from copper, and heat gently for some time ; wash, dry, and
ignite the precipitate, and weigh it as bismuth trioxide,
taking care to detach the dried carbonate as completely as
possible from the paper before incineration. Boil the solu-

tion containing the copper, add caustic soda, and continue the boiling until the solution is free from ammonia, filter, wash, dry, and ignite.

The filtrate from the original precipitate by sulphuretted hydrogen contains the iron, zinc, nickel, cobalt, and man-·ganese. It is concentrated slightly, mixed with nitric acid, and boiled until the iron is peroxidised, allowed to cool, mixed with a large quantity of strong solution of ammonium chloride, and then, drop by drop, with ammonium carbonate, until the fluid is just turbid (it must not show the least trace of distinct precipitate). Heat to boiling, and maintain the fluid in ebullition until the carbonic acid has been expelled. Allow the precipitate to settle, add one drop of ammonia to the clear liquid, and, if no precipitation ensues, a few cubic centimetres of ammonia : filter, and wash the precipitate with water containing a little ammonium chloride. Dry, ignite, and weigh the ferric oxide. The filtrate contains the manganese, nickel, cobalt, and zinc. Concentrate to a small bulk, and then add a strong solution of sodium acetate and acetic acid until the fluid is distinctly acid, heat to boiling, and whilst hot pass a rapid current of sulphuretted hydrogen into the liquid. The zinc, nickel, and cobalt are precipitated ; the manganese remains in solution. The precipitate is thrown upon a filter and washed with water containing sulphuretted hydrogen.

The manganese in the filtrate is determined by boiling the liquid to expel the sulphuretted hydrogen, adding caustic soda until it is nearly neutral, and then a few drops of bromine, and warning until the manganese separates out as binoxide. This is filtered, well washed, and converted into protosesquioxide (Mn_3O_4) by ignition. The filtrate will contain any lime and magnesia derived from the gangue ; these earths may be separated in the usual manner by ammonium oxalate and sodium phosphate.

The mixed sulphides of zinc, nickel, and cobalt on the filter are dried, transferred to a small beaker, the filter burned, and the ash added, and the whole dissolved in dilute aqua regia; solution of caustic potash is added until the solution is slightly alkaline, then acetic acid until it is distinctly acid, and lastly a strong solution of potassium nitrite. Allow the solution to stand for at least 24 hours, take out a small portion of the clear liquid, and add to it a few more drops of potassium nitrite; if after the lapse of a couple of hours no further precipitation ensues, the separation of the cobalt is complete. Pass both portions of the solution through a small filter, wash and dry the yellow crystalline potassium-cobalt nitrite, transfer it to a small crucible, incinerate the filter, add the ash, moisten the whole with strong sulphuric acid, expel the excess by heat, and weigh the residue; it has the composition $2CoSO_4 + 3K_2SO_4$, and contains 18·0 per cent. of cobalt oxide.

The filtrate containing the zinc and nickel is boiled, carbonate of soda added, and the metals separated as in No. XIV. Part II.*

XVII. Iron Pyrites

may be analysed by the methods adopted for copper pyrites.

XVIII. ' Kupfernickelstein '

is a mixture of sulphides of copper and nickel, containing arsenic, iron, cobalt, lead, &c. It is obtained as an intermediate product in the preparation of copper-nickel and German silver. It may be analysed by the processes described under copper pyrites.

XIX. Iron-Ores.

The ores of iron more commonly used for the extraction of the metal are the magnetic oxide, red and brown hæmatite, specular ore, spathic ore, and clay iron-stone. The following analyses show the characteristic features of these varieties.

---

* If the copper alone is to be determined in the ore, the above process is recommended whenever an accurate estimation is required.

	1	2	3	4	5	6
Ferric oxide .	70·23	94·23	90·05	2·75	2·72	0·40
Ferrous oxide .	29·65	—	—	48·12	40·77	45·86
Manganous oxide		0·23	0·88	0·83	—	0·96
Alumina . .		0·63	0·14	1·63	—	5·86
Lime . .		0·05	0·06	1·75	0·90	1·37
Magnesia . .		trace	0·20	2·29	0·72	1·85
Silica . .		4·90	0·92	1·62	10·10	10·88
Carbonic acid .		—	—	39·92	26·41	31·02
Phosphoric acid.		trace	0·09	0·54	—	0·21
Sulphuric acid .		0·09	} traces	—	—	trace
Iron Pyrites .		0·03		0·22	—	0·10
Water . .		0·56	9·22	0·45	1·00	1·08
Organic matter .	—	—	—	0·39	17·38	0·90
	99·88	100·72	100·76	100·51	100·00	100·29

1. Magnetic ore. Dannemora.
2. Red Hæmatite. Ulverstone.
3. Brown Hæmatite. Dean Forest.
4. Spathic ore. Westphalia.
5. Blackband ore. Scotland.
6. Clay iron-stone. Dudley.

General Method for the Complete Analysis of Iron Ores.—
The ore is carefully sampled, and an average portion of it is
reduced to a moderately fine powder, dried under the desiccator or at 100°, according to circumstances, and kept in a
well-corked tube or bottle.

Determination of the Moisture.—In ores containing carbonic acid, and organic matter, the water can only be
estimated by direct weighing. 2 or 3 grams of the powdered
ore are introduced into the bulb-tube (fig. 53), and ignited
in a slow current of air dried by the chloride of calcium tube:
the water is condensed in the weighed tube containing
calcium chloride. The rate at which the air passes through
the apparatus is seen in the small flask containing sulphuric
acid. After an hour's gentle ignition the tube is re-weighed ;
its increase in weight shows the amount of moisture present
in the sample.

Determination of the Carbonic Acid.—This is best effected
by means of the apparatus represented in fig. 31, p. 86.

Weigh out 1 or 2 grams of the ore into the flask A, and proceed as directed (see No. V. Part II.).

Determination of the Silica, Iron, Manganese, Sulphur, Phosphorus, Alumina, Lime, Magnesia, &c.—Introduce about 8 or 10 grams of the finely-powdered ore into a porcelain basin, and gently heat with concentrated hydrochloric acid, mixed with a little nitric acid, until the mineral is completely decomposed. Some varieties of ore—for example, blackband iron-stone—contain such an amount of organic matter that it is difficult to determine when they are completely dissolved. As a rule not more than 30 or 40 minutes will be required for complete decomposition. Some specimens of magnetic iron-stone and micaceous hæmatites dissolve with great slowness even in concentrated hydrochloric acid. In such cases it is better to heat the weighed portion of the finely-divided ore in a current of hydrogen or coal gas until water ceases to be evolved : the reduced iron will then readily dissolve in the acid. Fusion with acid sulphate of potassium also effects the decomposition of such ores. The solution is evaporated to dryness, and the residue drenched with strong hydrochloric acid, the liquid warmed, diluted with hot water, allowed to settle, and filtered into a ½-litre flask. The residue in the dish is again heated with a small quantity of hydrochloric acid, diluted with water, allowed to settle, and the clear liquid poured through the filter. This process is repeated until the silicious residue appears quite white and free from iron. It is then thrown on to the filter and washed thoroughly with hot water, dried, ignited, and weighed. It consists of gangue and separated silica. The amount of silica may be determined by boiling the weighed residue with a solution of sodium carbonate in a platinum dish, filtering and determining the weight of the gangue remaining : the difference gives the quantity of silica. It is sometimes required to determine the nature of the gangue : this is accomplished by the methods given in Nos. XI. & XII. Part II. Titanic acid is not an unfrequent constituent of iron-ores : when present it will be found partly in the silica separated by evaporation to dryness, partly in the hydrochloric acid solution. In order to determine its amount a large quantity of the ore is decomposed by hydrochloric acid, in the manner above described, evaporated to dryness, drenched with hydrochloric acid, and the residue filtered off and washed. The silicious matter is transferred to a platinum dish and repeatedly treated, in a 'draught-place' or in the open air, with hydrofluoric acid, and a little sulphuric acid;* the residue is fused with potassium-hydrogen sulphate, dissolved in a little cold water, and filtered, if necessary. Ammonia is added in *slight* excess, the solution boiled for some time, filtered, and the precipitated titanic acid washed, dried, ignited, and weighed. To the hydrochloric acid solution a few cubic centimetres of nitric acid are added, and the liquid is boiled, ammonia added, and the liquid again boiled. The precipitated oxide of iron and alumina carries down the rest of the titanic acid : these are filtered off, washed, dried, and transferred to a platinum dish, and fused with potassium-hydrogen sulphate. The mass is dissolved in a large quantity of cold water, neutralised with sodium carbonate, and solution of sodium thiosulphate added in slight excess. When the solution becomes nearly colourless, boil until all sulphur dioxide is expelled, filter, wash the precipitate with hot water, dry, and ignite it gently in a porcelain crucible. The residue contains all the titanic acid mixed with alumina. It is treated with strong sulphuric acid, filtered if necessary, diluted and boiled for some time, and the separated titanic acid filtered off and weighed.

The filtrate from the silica, &c., separated from the 10 grams of iron, is diluted to 500 cubic centimetres and well mixed by shaking.

Determination of the Sulphur.—Take out 100 cubic centi-

* If the sulphuric acid be omitted, a portion of the titanium will be lost by volatilisation as fluoride.

metres of the solution and evaporate nearly to dryness to expel the greater portion of the free acid, dilute with about 200 cubic centimetres of water, boil, and add one or two drops of barium chloride solution. After standing about 24 hours the barium sulphate is filtered off and weighed.

Determination of the Phosphoric and Arsenic Acids.—To 100 cubic centimetres of the solution, add a few cubic centimetres of a clear solution of molybdate of ammonium in nitric acid. [This solution is prepared by dissolving 10 grams of powdered ammonium molybdate in 40 c.c. of dilute ammonia (sp. gr. 0·96), and mixing the solution with 160 c.c. of dilute nitric acid (120 cubic centimetres of strong acid to 40 cubic centimetres of water). The mixture is heated to about 40° for some hours, and the clear liquid drawn off.] The mixture of iron salt and molybdate is kept in a warm place (not above 40°) for 24 hours, filtered, and washed, the precipitate treated with ammonia on the filter, and magnesia-mixture added to precipitate the dissolved phosphoric acid. After standing a few hours the magnesium ammonium phosphate is filtered, and treated in the usual manner. Many iron-ores contain notable quantities of arsenic, which is precipitated from the solution on adding molybdic acid. If qualitative analysis has shown the presence of arsenic, it must be removed before precipitating the phosphoric acid by transmitting a current of sulphuretted hydrogen through the 100 cubic centimetres of iron solution, filtering, heating the filtrate, if turbid, with a little nitric acid, and then adding the molybdic acid solution. The arsenic in the precipitate does not represent the entire amount in the ore, since a considerable portion is lost on the evaporation of the solution to dryness in order to render the silica insoluble. If it be desired to determine the amount of arsenic actually present, a larger quantity of the ore must be heated with aqua regia, filtered, and treated with sulphuretted hydrogen, and the arsenic determined in the precipitate (see No. XVIII.

Part II.) Any black residue left after oxidation may consist of copper oxide.

Determination of the Manganese, Alumina, Lime and Magnesia, Potash and Soda.—100 cubic centimetres of the solution are boiled with a little nitric acid, acid carbonate of ammonia added until the fluid is nearly neutral, and then to the *clear* red liquid, ammonium acetate in excess : boil for some time, until the precipitate settles on removing the lamp. Filter into a flask, wash with water containing a little ammonium acetate, dry, ignite, and weigh. The precipitate consists of ferric oxide, alumina, and phosphoric acid, and contains traces of silicic acid left in solution. The weighed substance is fused with acid-sulphate of potassium, the fused mass treated with hot water, and the silica filtered off and weighed. The amount of alumina is determined by subtracting the total quantity of ferric oxide, phosphoric acid, and silica from the original weight of the ignited precipitate. If it is required to determine the alumina directly, add tartaric acid to the solution, then ammonium chloride, and ammonium sulphide. The precipitate after standing a few hours is filtered off, and washed with water containing ammonium sulphide. Add sodium carbonate and nitre to the filtrate, evaporate to dryness, ignite, dissolve in dilute hydrochloric acid, filter if necessary, and precipitate the alumina by the addition of ammonium chloride and ammonia in slight excess, boil until the ammonia is expelled, and wash thoroughly with hot water. From the weight of this precipitate the amount of phosphoric acid is to be subtracted, since it is precipitated in the process together with the alumina.

The filtrate from the basic acetates contains the manganese, alkalies, and alkaline earths. A few drops of bromine are added, the solution is heated to 40 or 50°, and the flask tightly corked. In a few hours the whole of the manganese separates out as binoxide : it is filtered off, dried, ignited, and weighed as Mn_3O_4. Concentrate the filtrate, add ammonia and ammonium oxalate, and convert the precipitate into lime by ignition. The filtrate contains the magnesia and alkalies mixed with a large quantity of ammonium salts. It is evaporated to dryness, ignited to expel the ammoniacal salts, treated with a small quantity of water, about 1 gram of oxalic acid added, and the solution again evaporated to dryness and ignited. In this process the alkalies are left as carbonates, and the magnesium chloride is converted into magnesia. The residue is treated with a little water, and the magnesia filtered off, dried, and weighed. The carbonates in solution are converted into chlorides, weighed, and separated as in No. IV. Part II. If great accuracy is required, the filtrate containing the excess of platinum is reduced by means of hydrogen, filtered from the precipitated metal, and the small quantity of magnesia usually present in solution precipitated by the addition of ammonia and a drop or two of sodium phosphate. (See Analyses of Plant-ashes.)

Determination of the Iron by Solution of Potassium Bichromate.—When a solution of potassium bichromate is added to a liquid containing a ferrous salt and free hydrochloric acid, the iron is converted into a ferric salt in accordance with the reaction :

$$6FeCl_2 + K_2Cr_2O_7 + 14HCl = 3Fe_2Cl_6 + 2KCl + Cr_2Cl_6 + 7H_2O.$$

The final point of the reaction—that is, the point at which the whole of the iron is converted into ferric chloride—is ascertained by bringing a drop of the solution in contact with potassium ferricyanide, when no blue colouration will be produced. So long as the faintest trace of ferrous chloride remains, a drop of the liquid will colour the potassium ferricyanide. It is evident from the equation that 1 eq. or 294·42 parts of the bichromate will convert 6 eq. or 336 parts of iron to the state of a ferric salt. By dissolving 4·907 grams of pure dry potassium bichromate in a litre of water, a solution is obtained of which 1 cubic centimetre is equivalent to ·0056 gram of iron.

The potassium bichromate is purified by recrystallisation, dried between blotting paper, fused, and roughly powdered. About 5 grams of the salt are then weighed out into a litre flask, and diluted with so much water that each cubic centimetre contains 0·004907 gram of the bichromate. Suppos-

FIG. 58.

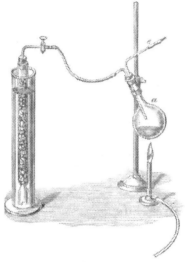

ing that 5·073 grams have been weighed out, this would require 1033·8 cubic centimetres, since

$$4·907 \; : \; 1000 \; :: \; 5·073 \; : \; 1033·8.$$

The litre flask containing the salt is filled up to the containing-mark, and the 33·8 cubic centimetres added from a burette. If there is not sufficient space between the mark

and stopper of the flask, the 33·8 cubic centimetres are poured
into the dry bottle in which the solution is to be preserved,
and the 1,000 cubic centimetres of bichromate solution added;
the liquid is well shaken, the litre flask refilled with it, and
the liquid poured back : if this process is repeated two or
three times the solution will be of uniform strength.   It is
advisable, however, to test the strength of the solution by
direct experiment.   For this purpose about 0·2 gram of fine
pianoforte wire is weighed out with the greatest care, and dissolved in a few cubic centimetres of pure hydrochloric acid,
in the flask *a* represented in fig. 58.   During the solution a
slow current of carbonic acid is passed through the apparatus;
the exit tube *e* is furnished with a little valve, made by
cutting a short slit in a small piece of caoutchouc tubing,
slipping the one end over the tube and stopping the other
by means of a glass rod.*   This little valve opens by inward
pressure only ; as soon as the pressure is applied outwardly,
the sides of the slit are pressed together and effectually
prevent the entrance of air.   In this manner any chance of
the ferrous solution within the flask oxidising is prevented.
Whilst the iron is dissolving, make a solution of potassium
ferricyanide by dissolving a minute portion of the salt in
about 15 cubic centimetres of water in a test-glass.   The
solution of the ferricyanide should be very dilute, otherwise it gives a reddish precipitate towards the completion of
the assay.   Spread a few *small* drops over a plate or porcelain slab by means of a glass rod, and fill up the burette with
the solution of bichromate.   When all the iron is dissolved,
the solution is boiled for a minute, and mixed with a quantity
of cold water, and the bichromate is added to it with
constant stirring.   The solution in the flask changes rapidly

---

* It is convenient, when a number of such estimations have to be
made, to determine the weight of a certain length of the piano-wire, and
to cut off a portion when needed equivalent to 0·2 gram.   The amount
taken must of course be controlled by weighing the wire.   A millimetre scale fastened on to the balance-table will be found very useful
for this and similar purposes.

in colour and becomes dark green. A brown colour indicates that a deficiency of hydrochloric acid is present. From time to time a drop of the solution is brought from the flask in contact with the ferricyanide upon the slab. When the intensity of the blue produced begins to diminish, the bichromate solution must be more slowly added. The mixed drops soon acquire a greenish tint, and when the last trace of this colour disappears the reaction is finished. A slight correction requires to be made on the quantity of iron taken, on account of the impurity present in it. If pianoforte wire be used, it may be assumed, without sensible error, that it contains 99·7 per cent. of iron. *

Example.—A solution of bichromate was made in accordance with the directions given: 5·073 grams of the recrystallised and fused salt being dissolved in 1033·8 cubic centimetres of water. 0·2097 gram of piano-wire was then dissolved in hydrochloric acid. This solution required 38·0 cubic centimetres of bichromate solution before the blue colour disappeared: 38·0 cubic centimetres bichromate are therefore equal to (0·2097 × ·997) = ·2091 gram of iron, or 1 cubic centimetre of bichromate is equivalent to 0·0055 gram of iron.

25 cubic centimetres of the solution of the iron ore to be tested are measured off into the flask, a few cubic centimetres of hydrochloric acid added, and two or three small pieces of pure zinc. The flask is connected with the carbonic acid apparatus, and the evolution of the hydrogen promoted by a gentle heat. When the solution has become nearly colourless, or possesses only a faint tinge of green, a minute drop is withdrawn and tested with potassium thiocyanate, water is added, and the bichromate solution is poured in until the iron is completely converted into ferric chloride.

* The bichromate solution may also be standardised by means of ferrous sulphate precipitated by alcohol (p. 150): the salt thus purified may be preserved in a stoppered bottle without experiencing the least change.

The experiment is repeated with a second portion of 25 cubic centimetres of the original solution.

In ores containing a mixture of ferrous and ferric oxides, as magnetic or spathose iron-stones, the amount of the former oxide is determined by dissolving from 1 to 3 grams of the sample, according to its supposed richness, in hydrochloric acid, in a stream of carbonic acid, adding water, and titrating with bichromate solution according to the method just given.

XX. TITANIFEROUS IRON-ORE (ILMENITE).

Ilmenite is a naturally occurring ferrous titanate ($FeTiO_3$), but it is seldom met with quite pure ; it usually contains ferric oxide, manganous oxide, magnesia, alumina, silica, &c. Many iron-sands contain considerable quantities of ilmenite, associated with magnetic oxide of iron. As such ores are occasionally used in the manufacture of iron, it may be desirable to give a method for their analysis.

The weighed portion of the mineral, in a state of fine powder, is fused with about eight times its weight of acid sodium sulphate, and on cooling the fused mass is digested with cold water. The insoluble matter is filtered off, washed, dried, and weighed : it is usually free from titanic acid. It should, however, be treated by the method given in No. XIX., p. 218. Dilute the filtrate considerably, add a little nitric acid to it, and boil for some time to precipitate the titanic acid. Filter the precipitate, wash, dry, and ignite it, with the addition of a fragment of ammonium carbonate, to ensure the volatilisation of sulphuric acid, which the precipitate is apt to carry down with it. Titanic acid is slightly hygroscopic: it must be weighed, therefore, as expeditiously as possible. The iron, magnesia, lime, and alumina remain in solution, and may be separated by the methods given in No. XIX., p. 217.

Titaniferous ores may also be decomposed by heating with sulphuric or hydrochloric acid, under pressure. A rapid and accurate method of estimating iron and titanium when present in solution together, after fusing the ore with acid

Q

sodium sulphate, and treating the fused mass with water, is founded on the behaviour of these substances, when reduced to their lowest state of oxidation, towards a solution of potassium permanganate. The solution thus obtained is diluted to a determinate amount, an aliquot portion withdrawn, and reduced by zinc and sulphuric acid in the apparatus represented in fig. 58 (p. 222). The iron and titanic acid are thus brought to the state of ferrous and titanous oxides, and their joint amount is determined by addition of standard permanganate solution until the rose-colour of the latter solution is permanent. A second aliquot portion is brought into the flask, and an apparatus for the disengagement of sulphuretted hydrogen or sulphur dioxide is substituted for the carbonic acid flask. The iron is thus reduced to the ferrous state, the titanic acid remains unchanged. The solution is boiled to expel the excess of the reducing agent, whereby a portion of the titanic acid is precipitated, filtered rapidly, and the solution is again titrated with permanganate. The amount of ferrous oxide is thus obtained : the difference between the titrations gives the amount of permanganate corresponding to the titanic acid present. The weight of the latter substance is found from the equation

$$Ti_2O_3 + O = 2TiO_2.$$

16 parts of oxygen are therefore equivalent to 164 parts of titanic oxide.

This process is also applicable to the analysis of rutile (native titanic oxide), sphene or titanite (a silico-titanate of calcium, $CaSiO_3.CaTiO_3$), &c.

XXI. Wrought and Cast Iron, and Steel.

Cast iron, in addition to chemically-combined carbon and graphite, contains variable quantities of silicium, phosphorus, sulphur, manganese, and copper ; and in very much smaller quantities, aluminium, chromium, titanium, zinc, nickel, cobalt, arsenic, antimony, tin, vanadium, magnesium, potassium, lithium, sodium, and nitrogen. The greater number of these substances are present in such very minute quantities, that their exact quantitative determination is a matter of great difficulty. It is very rarely necessary to attempt their estimation, since their presence in such small amount probably exercises little or no influence on the quality of the iron. Cast iron occurs in two leading varieties, viz. as *grey* and as *white* cast iron. The difference between the varieties is determined by the state of the carbon present in them. In grey cast iron the greater amount of the carbon is in the form of graphite, interspersed throughout the mass in a state of mechanical mixture; in the white variety the carbon is chemically combined with the iron. Intermediate varieties, in which grey and white cast iron are mixed together in varying proportions, are classed as *mottled* cast iron.

The following analyses of grey and white cast iron obtained from the same ore (spathic, containing about 42 per cent. iron) clearly show this characteristic difference :—

	Grey.	White.
Iron	90·584	93·183
Combined Carbon . . .	—	4·100
Graphite.	2·795	—
Silicon	4·414	0·230
Sulphur	0·039	0·030
Phosphorus	0·099	0·073
Manganese	1·837	2·370
Nickel and Cobalt . . .	traces	0·014
	99·768	100·000

Wrought iron approaches more nearly to the character of pure iron : it contains much less carbon than cast iron, melts at a higher temperature, and is heavier. It also contains much less silicon, but the amounts of sulphur and phosphorus are as variable as in cast iron. The subjoined analysis is of Swedish iron of excellent quality :—

Iron	99·863
Carbon	0·054
Silicon	0·028
Sulphur	0·055
Phosphorus	traces
	100·000

Steel is intermediate in properties and purity between cast iron and wrought iron : it differs, however, from these varieties by its remarkable and valuable property of becoming hardened by heating and sudden cooling.   In this state it is extremely brittle, and is without the characteristic fibre of malleable iron.   Its tenacity, however, is much greater than that of wrought iron.   Its average specific gravity is intermediate between that of cast iron and wrought iron. The following is an analysis of steel of medium quality :—

Carbon	0·501
Silicon	0·106
Sulphur	0·002
Phosphorus	0·096
Manganese	0·144
Iron (by difference) . . .	99·151
	100·000

The action of acids upon these varieties of iron is remarkable.   When heated gently with *strong* hydrochloric acid white cast iron is completely dissolved, whereas grey cast iron leaves a residue of graphite.   The combined carbon present in both varieties combines with the nascent hydrogen evolved by the action of the acid, giving rise to hydrocarbons of the C_nH_{2n} series, C_2H_4 . . . C_6H_{12}, which communicate a peculiar odour to the issuing gas.

The diluted acid at ordinary temperatures attacks cast iron but slowly : when dissolved by the aid of heat, white cast iron deposits a portion of its combined carbon ; this is soluble in potash, and on ignition leaves a black residue, containing silica.   The residue from grey cast iron, in addition to graphite, also contains this carbonaceous matter together with a black magnetic substance containing iron.   When dried this residue occasionally takes fire in contact with oxygen, and is converted into a mixture of ferric oxide and silica.

It is said that the action of acids upon steel varies with its hardness : soft steel is far more readily dissolved than hardened steel. Concentrated hydrochloric acid dissolves soft steel completely, but when diluted it leaves a larger amount of the black magnetic substance above mentioned than is yielded by malleable iron. It is worthy of note that diluted nitric acid completely dissolves this carbonaceous matter, forming a yellowish-brown coloured liquid, the intensity of the colour being proportional to the amount of combined carbon present. It is said that steel may be distinguished from cast and wrought iron by the action of hydrochloric acid of sp.gr. 1·134. With steel the acid occasions a rapid evolution of gas, which suddenly ceases in a short time (in about 20 seconds), whereas with cast or wrought iron the disengagement is continuous.

The complete analysis of iron, in addition to the estimation of the metal, necessitates the determination of the carbon, existing both in a state of combination and as graphite, of silicon, sulphur, manganese (zinc, cobalt, alumina, titanic acid), phosphorus, nitrogen, and admixed slag. The iron, however, is usually determined by difference.

Determination of the Total Carbon.—Among the many accurate methods which have been proposed for the estimation of the total carbon contained in iron, those of Wöhler, Weyl, and Ullgren are distinguished by reason of the ease and expedition with which they may be carried out. In all these processes the carbon is ultimately weighed as carbon dioxide.

(*a*) *Wöhler's Method.*—By burning the iron in a stream of oxygen. The iron must be previously reduced to the finest possible state of division, by filing with a hard file, and powdering in a large agate mortar. If the metal is very hard it is broken on a clean anvil, stamped to powder in the steel mortar, and passed through a fine sieve. From 3 to 6 grams of the metal (according to its supposed richness in carbon) are weighed out into a platinum boat, and brought

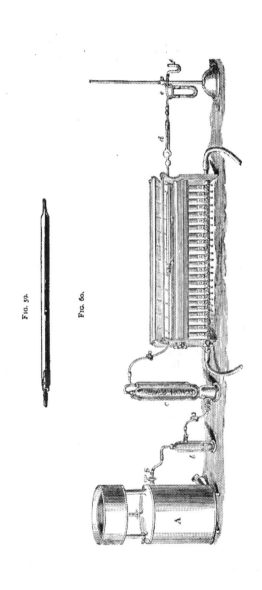

Fig. 59.

Fig. 60.

into a short piece of combustion tubing drawn out at one end to a narrow tube (fig. 59). The other end is closed with a caoutchouc cork, and is connected with a gasometer filled with oxygen. To the narrowed end of the combustion tube is attached a chloride of calcium tube, connected with a weighed U-tube filled ⅞ with soda-lime, and ⅛ with calcium chloride (fig. 60). A is a gasometer containing oxygen; by means of the cock s the amount of the issuing gas can be easily regulated. The rate of its passage can be seen in *b,* which contains solution of caustic potash, destined to absorb any traces of carbonic acid and chlorine which may be present in the oxygen. The cylinder *c* is partially filled with soda-lime with the same object: the upper half and the two U-tubes contain calcium chloride, by which the gas is thoroughly dried. The combustion tube containing the weighed amount of iron rests in a gas furnace; *d* is a chloride of calcium tube: it is placed merely as a precaution against moisture passing into the weighed U-tube, and need not be weighed. The little U-tube *f* contains one or two drops of concentrated sulphuric acid, sufficient to fill the bend; its object is to prevent the possibility of the calcium chloride in the weighed tube absorbing atmospheric moisture. The process of combustion needs no particular attention; if the heat is sufficiently powerful and the iron finely-divided, the whole of the carbon is converted into carbon dioxide. When the process is finished, *b* is detached from the gasometer, and a slow current of air is aspirated through the apparatus to expel the oxygen before the U-tube is again weighed.

(*b*) *Weyl's Method.**—In pulverising the iron for analysis, especially if very hard, there is considerable risk of mixing the sample with iron from the file, &c., employed. Weyl's method, by dispensing with the necessity of reducing the iron to powder, obviates this inconvenience. A piece of the

* This method appears to have been described so far back as 1857 by Binks, in a paper read before the Society of Arts.

iron weighing from 10 to 15 grams is suspended by a pair of pincettes provided with platinum points in a beaker containing dilute hydrochloric acid. Care must be taken that the platinum points in contact with the iron are not moistened with the acid, or its solvent action will be impeded from the separation of carbon between the points and the metal. Connect the upper portion of the pincettes with the wire of a positive pole of a single element of Bunsen's battery. To the wire of the negative pole is attached a slip of platinum foil, which is also immersed in the liquid. By regulating the distance between the foil and the metal, the strength of the current may be so modified that not a trace of ferric chloride is produced. The due regulation of the intensity is of the utmost importance, for if it is too strong the iron becomes passive, and chlorine is evolved from its surface, which brings about the oxidation of any separated carbon. This formation of chlorine is immediately rendered evident by the yellowish colour of the concentrated solution of iron as it falls away from the metal. When the operation succeeds, hydrogen only appears at the platinum foil, no gas being evolved from the positive pole (the iron). In 12 or 15 hours the whole of the iron immersed will be dissolved : the separated carbon retains the shape of the metal. The undissolved portion is detached from the spongy mass of carbon, dried, and weighed. Its weight subtracted from that of the iron originally taken gives the amount employed for the carbon determination.* The carbon is washed slightly, and thrown into a short piece of combustion tube containing a plug of ignited asbestos. This acts as a filter and retains the carbon : care must of course be taken that the plug is sufficiently compact to prevent any particles of carbon passing through with the filtrate. The tube is gently heated

---

* Instead of suspending the iron by the pincettes as above described, it may be broken into several pieces, and supported on a small platinum tray pierced with a number of holes. In this way the whole of the iron may be dissolved.

and a current of air aspirated through it. When perfectly freed from moisture, the plug together with the carbonaceous residue, is drawn by means of a bent wire into the middle of the tube, and is mixed with a small quantity of copper oxide by the aid of the wire. The combustion tube is heated in a gas furnace, and the carbon dioxide collected in a weighed soda-lime tube as described above. To ensure perfect combustion it is advisable to heat the carbon and copper oxide in a stream of oxygen : for this purpose the apparatus represented in fig. 60 may be employed. Instead of burning it, the separated carbon may be treated as in the next method.

(c) Ullgren's Method.—It is necessary for this method that the sample should be in a state of coarse powder. Grey cast iron is preferably taken in the form of borings : white cast iron should be coarsely powdered. Dissolve 10 grams of copper sulphate in about 50 cubic centimetres of water, and into this solution, contained in a small beaker, weigh out about 2 grams of the iron. Heat gently and with constant stirring until the iron is completely dissolved ; allow the solid portions to settle and pour away as much as possible of the clear fluid. Rinse the solid particles with any adhering copper solution into the small flask A (fig. 31), which should have a capacity of about 150 cubic centimetres. Care must be taken not to employ too large a quantity of wash water in rinsing the reduced copper and carbon into the flask : the total fluid in the flask should not exceed 25 cubic centimetres. Add to the flask 40 cubic centimetres of concentrated sulphuric acid : if more than 25 cubic centimetres of wash water have been used, proportionally more acid must be employed. Allow the mixture to cool, add about 8 grams of chromic acid, and connect the flask with the system of tubes represented in fig. 31. Heat the liquid gradually, and regulate the flame so as to maintain a regular evolution of gas : as it slackens, increase the heat, until white fumes make their appearance in the body of the flask. Remove the lamp

and aspirate a slow current of air through the apparatus (3 or 4 litres). Weigh the soda-lime tube and again aspirate air, in order to determine if the increase of weight due to the absorption of the carbon dioxide is constant. The principle of the method is evident. In contact with the sulphuric and chromic acids the carbon separated on dissolving the iron is converted into carbon dioxide.

Note on the preparation of Chromic Acid.—300 grams of coarsely-powdered commercial potassium bichromate are warmed with 500 cubic centimetres of water and 420 cubic centimetres of sulphuric acid. When the salt is dissolved the solution is allowed to stand ten or twelve hours, when the acid potassium sulphate separates out. The mother liquor is decanted, and the crystals allowed to drain for an hour. The solution is heated to 80° or 90°, and gradually mixed with 150 cubic centimetres of sulphuric acid, and then with the same quantity of water, when the precipitated chromic acid will be redissolved. The solution is concentrated in a porcelain basin until small spicular crystals appear on the surface of the liquid : after standing a few hours an abundant crop of chromic acid crystals will be obtained. The mother liquor on further evaporation will yield a fresh quantity of crystals. The chromic acid thus obtained may be drained by means of the filter pump, and dried on a porous tile placed beneath a bell jar: it is very hygroscopic and must be kept in a well-stoppered bottle. The crystals are not quite pure, as they contain small quantities of potash and sulphuric acid, but the presence of these substances in no way interferes with their employment in the above method.

Instead of absorbing the liberated carbonic acid by means of soda-lime, Ullgren prefers to use potash-pumice. This is prepared as follows :—1 part of caustic potash is dissolved in 3 or 4 parts of water, the solution is heated to 100° in an iron pot, and a quantity of pumice, in pieces somewhat less than the size of a pea, is added until the mass becomes nearly dry. Whilst still hot it is transferred to a wide-mouthed stoppered bottle, and briskly shaken until, on cooling, the small pieces no longer adhere to each other.

Determination of the Graphite.—The above methods, it will be observed, determine merely the total amount of carbon, and give us no information respecting the proportion present as graphite and as combined carbon. In order to determine the graphite, 3 to 5 grams of the sample are dissolved in moderately-concentrated hydrochloric acid at a gentle heat (comp. p. 228). The combined carbon combines with the nascent hydrogen, and is evolved together with the excess of this gas : the graphite remains undissolved. Filter the residue through asbestos, contained in a short piece of combustion tube (fig. 59, p. 230), wash it with hot water, then with potash, alcohol, and a little ether. Dry it, mix it with a little copper oxide, and burn it in a stream of oxygen as directed on p. 233 (see fig. 60). On deducting the weight of the graphite thus obtained from the total amount of the carbon, the difference gives the quantity of combined carbon. This process is the most uniformly accurate, but for technical purposes it may be thus simplified. Treat the weighed sample of iron with dilute hydrochloric acid, and when the metal is nearly dissolved, add a large quantity of strong hydrochloric acid. The insoluble matter is collected on a weighed filter, washed with hot water, with dilute hydrochloric acid to remove iron, and with potash to remove silica. In washing with the alkali there is occasionally a brisk evolution of hydrogen, owing to the oxidation of the silicon to silicic acid. The insoluble matter is lastly washed with alcohol and ether to remove any traces of adhering hydrocarbons, dried at 120° and weighed. The filter is thrown into a small platinum crucible, and incinerated : the weight of the residue, less that of the filter-ash, gives the amount of silica (and titanic acid) mixed with the graphite.

Estimation of Combined Carbon in Steel and Wrought Iron.

Eggertz's Method.—In the case of metallic iron containing but a small proportion of combined carbon, this method is readily applicable. When such iron is dissolved in nitric acid, the solution becomes coloured more or less brown in proportion to the amount of carbon present (comp. p. 229). By comparing the depth of this coloration with that of a standard tint, equivalent to a known quantity of carbon, the

determination of the combined carbon in a sample of steel
or wrought iron may be made in a very short time, and with
great accuracy. The process is therefore especially applicable in steel works, where such determinations are of frequent
occurrence. It is thus conducted. Two thin test-tubes,
made of the same glass, are divided into 0·5 cubic centimetre
by means of a pipette, the graduation being scratched on the
side by a diamond. A piece of steel weighing not less than
100 grams, supposed to contain from 0·7 to 1·0 per cent. of
carbon, is finely powdered, and the amount of carbon it contains determined by one of the methods above described.
As this sample is to serve as the standard of comparison,
the amount of carbon which it contains must be estimated
with the utmost possible accuracy : it is advisable therefore
to make several determinations, and to take the mean of the
results. Supposing, for the sake of illustration, that the
sample contains 0·75 per cent. carbon. One decigram of
the steel or wrought iron to be tested, and one decigram of
the standard steel (containing 0·75 per cent. carbon), are
weighed out with the greatest accuracy on small tared watch-
glasses. The samples are brought into thin dry test-tubes,
and covered with about 2 cubic centimetres of dilute nitric
acid (sp. gr. 1·2), perfectly free from chlorine. In a few
minutes the greater portion of the metals will be dissolved.
The tubes are placed in a beaker containing water at 80°,
which is maintained at this temperature until all action on
the metals is at an end. In the case of steels this ceases in
about two hours. Allow the tubes to cool, and decant the
clear supernatant liquid from the undissolved matter into the
graduated test-tubes ; the solution of the standard steel is
poured into the one tube, that of the iron into the other.
To each portion of residue add two or three drops of nitric
acid, and heat gently over the lamp : if no further evolution
of gas occurs, the insoluble matter consists merely of silica or
graphite. Allow these solutions to cool, and add them to
the contents of the graduated tubes. Dilute the solution

of the standard steel until it exactly measures 7·5 cubic centimetres : each cubic centimetre is equivalent therefore to o·oooi gram of carbon. Now add water, drop by drop, to the liquid contained in the other tube, agitate the mixture, and compare the tint with that of the liquid in the standard tube. If the tints are equal read off the volume of the liquid : each cubic centimetre represents one tenth per cent. of combined carbon. Thus, supposing that on diluting the liquid to be tested to 4·5 cubic centimetres it gave a depth of colour equivalent to that of the standard solution, then the sample contains o·45 per cent. carbon. It must be remembered that this method is strictly comparative : to ensure accuracy the circumstances of temperature, time, amount of acid used, &c., must be as nearly as possible identical. The normal solution must be prepared afresh for each series of comparisons, as it gradually becomes paler on keeping, especially if exposed to light.

When such determinations of carbon in steel or wrought iron are of frequent occurrence, as in iron works, it is more convenient to prepare a number of tubes containing the brown solution, corresponding to different percentages of carbon. The coloured solutions may be readily obtained by digesting roasted coffee in dilute spirit : if kept in the dark when not in use they maintain their intensity of colour unimpaired for a long time. Fig. 61 represents a wooden frame containing the tubes : these are about ⅜ of an inch in diameter, and about 3½ to 4 inches long. The tubes after the introduction of the properly-diluted solutions are hermetically sealed. The liquid in the tube placed in the second hole in the rack is made to correspond to the tint of a solution of 1 gram of steel containing o·o2 per cent. of combined carbon in 15 cubic centimetres of nitric acid of sp. gr. 1·20. The solution in the tube in the fourth hole corresponds to that of the same quantity of iron containing o·o4 per cent. carbon ; and so on in regular succession, each tube increasing in value by 0·02 per cent. The last tube is equivalent therefore to 0·3 per cent. The process is thus conducted:

FIG. 61.

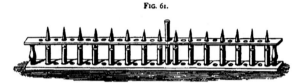

1 gram of the iron or steel to be tested, in the state of fine powder, is weighed out into a large test-tube, and digested at 70° or 80° with 10 cubic centimetres of the dilute nitric acid, free from chlorine, for about half an hour. The solution is quickly cooled by immersing the tube in cold water, and is filtered, without the residue being disturbed, through a *small dry* filter into a test-tube of the same size and made of the same glass as those containing the standard solutions. The insoluble matter is then treated with 5 cubic centimetres of the nitric acid, heated gently, and the solution added to the main portion. The entire solution is mixed by shaking, and its colour compared : the holes in the stand allow the tube to be placed side by side with the standard solution : the number affixed to the tube with which it corresponds in colour indicates the percentage amount of carbon in the sample. If the steel or wrought iron contains more than 0·3 per cent. of carbon, 0·5 gram is taken, or the solution is diluted with an equal volume of water, shaken, and half of it poured away. The comparison is assisted by placing a white screen of paper behind the tubes. In this manner a carbon determination may be made to within ·01 per cent.

Estimation of the Sulphur.—When iron containing sulphur is treated with hydrochloric acid, the whole of the sulphur is evolved as sulphuretted hydrogen. By passing the sulphuretted hydrogen into a solution of bromine in hydrochloric acid, it is completely absorbed and converted into sulphuric acid ; by converting the sulphuric acid into barium sulphate, the amount of sulphur may be readily determined. The apparatus required for the purpose is seen in fig. 62. The flask of 150 cubic centimetres capacity is fitted with a caoutchouc cork, which has previously been boiled in dilute caustic soda to remove the sulphur contained in it: the cork is pierced with two holes, into one of which fits a straight piece of tubing, curved at its lower extremity: near

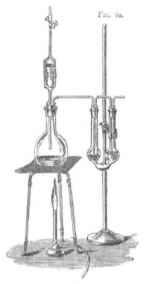

FIG. 62.

the upper end is a bulb filled with dilute hydrochloric acid : the tube is closed by means of a small piece of clean caoutchouc tubing and a clamp. The second hole of the cork contains a tube bent at right angles, to which is adapted a small U-tube, containing the solution of bromine. It will be seen by the arrangement of the glass tubes that the gas passes twice through the liquid contained in the U-tube.

Into the flask weigh out 10 or 15 grams of the finely-divided iron, fill the bulb with moderately diluted hydrochloric acid, insert the cork, open the clamp, and by aspiration at the exit-tube of the absorption apparatus bring the acid into the flask. Heat the flask gently, and introduce fresh acid from time to time until the iron is completely dissolved. Connect the U-tube with an aspirator, and gently draw a current of air through the apparatus to remove the last traces of sulphuretted hydrogen in the flask. Transfer the liquid from the U-tube to a beaker, boil to expel excess of bromine, and add barium chloride.

It is always advisable to test the solution of ferrous chloride for sulphuric acid by concentrating it, and adding one or two drops of barium chloride : if any precipitate is thus obtained it is to be filtered off, washed, dried, and weighed. The insoluble residue should also be tested for sulphur by fusing it with nitre and sodium carbonate, dissolving in water, and testing the acidified solution with barium chloride. Usually, however, the residue will be found to be quite free from this substance. A very convenient method of converting the sulphur in iron to the state of sulphuric acid consists in absorbing the sulphuretted hydrogen in a dilute solution of potash or soda free from sulphate. Decant the alkaline solution into a beaker, add a few drops of bromine free from sulphuric acid, heat gently, acidulate with hydrochloric acid, boil, add a few drops of barium chloride, and after standing about twenty-four hours filter off the precipitated barium sulphate.

The sulphuretted hydrogen may also be absorbed by a dilute solution of a cadmium salt : the cadmium sulphide possesses the advantage that it can be dried at 100° without alteration.

When a number of estimations of sulphur in iron have to be made, it is convenient to employ a volumetric method to determine the sulphuretted hydrogen evolved. Both the following plans will be found to give concordant results. The U-tube (fig. 62) is filled with a solution of caustic soda

(1 of soda to 5 of water) to absorb the sulphuretted hydrogen. The operation of dissolving the iron is made exactly as in the foregoing methods. When the action is at an end, pour the contents of the U-tube into a large beaker, dilute with about 200 cubic centimetres of boiled water, acidify with hydrochloric acid, add a little starch paste, and add standard iodine solution until the solution is turned blue.

Or wash the alkaline solution into a small beaker containing water, to which a measured quantity, say 10 cubic centimetres, of deci-normal arsenious acid solution is added. Add hydrochloric acid to distinct acid reaction, set aside for a few hours in a warm place, dilute to a determinate quantity, say 100 cubic centimetres, filter through a dry filter, withdraw 25 cubic centimetres of the filtrate, neutralise with sodium bicarbonate, add a small quantity of starch liquor, and a dilute solution of iodine from a burette until the blue colour is permanent. The determination is repeated with a second portion of 25 cubic centimetres. This method is based upon the following reaction between the arsenious acid solution and the sulphuretted hydrogen :

$$As_2O_3 + 3H_2S = As_2S_3 + 3H_2O.$$

1 cubic centimetre of deci-normal arsenious acid solution = 0·0024 gram sulphur. With proper care this method affords very accurate results.

Determination of Nitrogen.—From the experiments of Schafhäutl, Despretz, Boussingault, Frémy, and others, it appears that nitrogen is an invariable constituent of cast and wrought iron, and steel ; some of these authorities are of opinion that when present in estimable proportion it exerts a marked influence upon certain of the physical properties of the metal. The iron is said to be rendered white and brittle and less liable to change in the air or water (Despretz ; Buff). According to Frémy the nitrogen exists in the iron in two conditions, since when the metal is dissolved

R

in hydrochloric acid, a portion is converted into ammonia, whilst another portion remains in combination with the carbonaceous residue. On the other hand, Erdmann, Stahlschmidt, Stuart, and Baker assert that the quantity of nitrogen usually present in iron is too minute to have the least influence upon the metal. The following methods of determining the nitrogen present in iron are due to Ullgren.[*]

Determination of the Nitrogen which forms Ammonia on dissolving the Iron in Hydrochloric Acid.—About 2 grams of the finely-divided iron are treated, in a flask provided with a bent tube, with a solution of 10 grams crystallised copper sulphate, and 6 grams fused sodium chloride. When the iron is dissolved, add excess of boiled milk of lime, boil for some time, and determine the evolved ammonia as in No. VIII. p. 94.

The quantity of the ammonia in the distillate, if very minute, may be determined by Nessler's method (see Water Analysis).

Determination of the Nitrogen present in the Carbonaceous Residue.—By combustion with mercuric sulphate, and measurement of the evolved nitrogen. The apparatus employed for the purpose is seen in fig. 63. A is a piece of combustion tubing about 30 centimetres long ; it is filled as far as *b* with 12 grams of dry magnesite or bicarbonate of soda. From *b* to *c* is placed the mixture of about 0·1 gram of the carbonaceous residue (dried at 120-130°), with from 3 to 4 grams of the mercuric sulphate. The mixture is made in a small glazed porcelain or agate mortar, which is rinsed with a fresh portion of the sulphate after the introduction of the mixture. From *c* to *d* is a layer of coarsely-powdered pumice, previously mixed with mercuric sulphate[†] and a little water, and dried : its object is to prevent the possible evolution of carbon monoxide. The remainder of the tube is

[*] Ann. d. Chem. u. Pharm. cxxiv. 70, cxxv. 40.
[†] Preparation of the mercuric sulphate. This salt is obtained by gradually adding 4 parts of mercury to 5 parts of strong boiling sulphuric acid, and heating the mixture until sulphur dioxide ceases to be evolved, and the whole is converted into a dry saline mass.

filled with fragments of pumice soaked in a concentrated solution of potassium bichromate, and allowed to drain: its object is to absorb the sulphur dioxide which is disengaged. The various mixtures are separated by plugs of recently-ignited

FIG. 63.

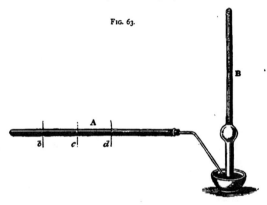

asbestos. The tube is fitted with a caoutchouc cork and bent tube. The vessel B is designed to collect and measure the evolved nitrogen. The narrow portion holds about 20 cubic centimetres, and is graduated into $\frac{1}{10}$ths of a cubic centimetre: the bulb holds about 40 cubic centimetres, and the lower portion from 20 to 30 cubic centimetres. Fill the tube with mercury and invert it in the trough. By means of a pipette pass up potash solution (1 part potassium hydrate and 2 of water) until the bulb is nearly filled (to within 10 cubic centimetres), and then add 15 cubic centimetres of a clear and saturated solution of tannic acid. Now gently heat the magnesite or bicarbonate of soda at the extreme end of the tube, and gradually heat the tube until about half the substance has been decomposed: the air within the apparatus is thus expelled. Bring the end of the delivery-tube under B, heat from *b* to *c* very gently to remove any moisture present, then heat from *c* to *d* gradually; and when

this portion of the tube is nearly red hot, heat the part
b c to a strong red heat. Heat from *b* to *d* until no more gas is
evolved, and sweep out the gases within the tube by heating
the undecomposed portion of magnesite. Close the end of
B with the thumb, and transfer the tube to a vessel of water :
the mercury and potash will be replaced by water ; adjust
the levels of the liquid within and without the tube, and read
off the volume of nitrogen, making the necessary corrections
for tension of aqueous vapour, temperature, and pressure,
and calculate the weight of the nitrogen found.

*Determination of the Silicon, Iron, Manganese, Cobalt,
Nickel, Zinc, Alumina, Titanic Acid, Alkaline Earths,
and Alkalies.*[*]—Weigh out about 10 grams of the finely-
divided iron into a porcelain basin, cover it with a large
watch-glass, and dissolve it in moderately dilute hydrochloric
acid, add a few drops of sulphuric acid, and evaporate to
dryness on the water-bath, heating until the mass no longer
smells of hydrochloric acid. Moisten the dried salt with
hydrochloric acid, heat on the water-bath, add water, filter into
a capacious porcelain basin, wash and dry the residue. Set it
aside and label it ' Pp. I.' Heat the filtrate with nitric acid,
boil, add water until the liquid measures at least 1,500 cubic
centimetres, and gradually add ammonium carbonate in dilute
solution until the fluid just loses its transparency (it must not,
however, show any sign of distinct precipitate). Heat to
boiling, and maintain the liquid in ebullition until the carbonic acid is expelled. If the solution is not too concentrated the ferric hydrate separates rapidly. Add now a few
drops of ammonia, filter, and wash the precipitate with water
containing a little ammonium chloride. The only condition
necessary to ensure success is the proper dilution of the
liquid : for the quantity of iron taken it should not measure
less than $1\frac{1}{2}$ litre. Dry the washed precipitate, set it aside,
and label it ' Pp. II.'

[*] Compare Lippert, Zeits. f. anal. Chemie, ii. 39.

Add ammonia in slight excess to the filtrate from ' Pp. II.,' and boil until the free ammonia is nearly expelled, filter, and, without washing, redissolve in hydrochloric acid, and again precipitate with ammonia. Filter, wash and dry the precipitate : call it 'Pp. III.'

Mix the two filtrates, acidify with hydrochloric acid, concentrate in a porcelain basin, transfer to a small flask, add ammonia and ammonium sulphide, cork the flask, and let the liquid stand in a warm place for at least twenty-four hours. Decant the clear fluid through a filter, and wash it with water containing a little ammonium sulphide ; allow it to drain as far as possible, spread the filter on a glass plate, and wash off the precipitate into a small flask ; add acetic acid, and cork the flask. Label the flask ' Pp. IV.'

Transfer the filtrate to a porcelain basin, evaporate, add nitric acid, and heat until the ammonium chloride is decomposed. Evaporate to dryness, dissolve in a little water, filter if necessary, add ammonium oxalate, and, after standing twenty-four hours, filter off the calcium oxalate, wash it and convert it into lime by ignition. Evaporate the filtrate to dryness in a platinum basin, ignite at first gently and then to a red heat. Treat with water, and filter off the magnesia and weigh it. The solution of the alkalies is acidified with hydrochloric acid, evaporated to dryness, the mixed chlorides weighed, and separated as in No. IV. Part. II.

The residue marked ' Pp. I. ' contains the graphite, silica, a portion of the phosphorus as phosphide of iron, titanic acid, barium sulphate. Fuse it with sodium carbonate and a little nitre, soften the mass with water, add hydrochloric acid in excess, evaporate to dryness and separate the silica. Weigh it, and treat it in a platinum basin with a moderately-concentrated solution of sodium carbonate : if it dissolves completely, it is pure ; if a weighable amount remains, determine its quantity and test it for titanic acid and barium sulphate. To the filtrate from the silica, add ammonia, filter off any precipitate which forms, redissolve it in hydrochloric acid, and again add ammonia. Filter and dry, and add the precipitate which forms to 'Pp. III.' To the filtrate add ammonium sulphide, and add any precipitate which separates out to 'Pp. IV.' Test the filtrate for the alkaline earths in the manner above directed, and if their amount admits of determination, weigh them.

'Pp. II.' and 'Pp. III.' contain the ferric oxide and alumina, and the small quantity of titanic acid which may have passed into the hydrochloric acid solution. The mixed precipitates are placed in several porcelain or platinum boats, and strongly ignited in a glass tube in a current of dry hydrogen until the formation of water ceases. Allow the reduced metal to cool in the current of the gas, and treat the mixture with dilute nitric acid (1 part of acid to 30 of water), until the iron is dissolved : filter the liquid into a litre flask, dilute to 1,000 cubic centimetres, and determine the iron by precipitating an aliquot portion by means of ammonia. This method is better than that of determining the iron in a fresh portion of the sample, unless the sample is perfectly homogeneous. The residue left after treatment with dilute nitric acid is fused with acid-potassium sulphate, digested with water, and any residual silica filtered off and weighed. The filtrate is boiled for some time, and if any titanic acid separates out, it is filtered off and weighed. Add ammonia and boil : dissolve the precipitate in hydrochloric acid, transfer the liquid to a small test-tube, add a little tartaric acid, then ammonia, and ammonium sulphide. Allow the liquid to stand until any iron sulphide which forms has completely settled, filter, redissolve in hydrochloric acid, boil with a few drops of nitric acid, add ammonia, and filter off the ferric hydrate ; dry, ignite, and weigh. To the yellow filtrate containing the alumina, add a little pure sodium carbonate and nitre, evaporate to dryness, and ignite until the mass is completely white. Rinse the residue into a beaker, dissolve in hydrochloric acid, and add excess of ammonia. Filter off the precipitate, wash, dry, and weigh it.

Mix the filtrate with a few drops of magnesium sulphate solution : if a precipitate of magnesium-ammonium phosphate forms after standing, calculate the weighed precipitate as $AlPO_4$. If no precipitate is formed, the amount of phosphoric acid, determined as under, is to be subtracted from the weighed precipitate. The remainder gives the amount of alumina. This precipitate will also contain any chromium present in the iron. Its amount is in general exceedingly minute. In case the quantity happens to be more considerable than usual, fuse the mixed oxides with a little sodium carbonate and nitre, dissolve in water, add a small quantity of ammonium nitrate, evaporate nearly to dryness, add a little water, filter, reduce the alkaline chromate with sulphurous acid, and precipitate the chromic oxide with ammonia, boil for some time, filter, dry, and weigh. The difference between the weight of the oxide thus obtained and that of the original precipitate gives the alumina.

'Pp. IV.' consists principally of manganese sulphide : it may also contain zinc, copper, nickel, and cobalt. By digestion with acetic acid the greater portion of the manganese sulphide will have been dissolved : filter, spread the filter containing the residue on a glass plate, and wash the precipitate into a small beaker containing sulphuretted hydrogen water acidulated with hydrochloric acid. Allow the liquid to stand for a short time : the zinc and the remainder of the manganese pass into solution ; the copper (which is estimated as under), cobalt, and nickel remain undissolved. Filter, evaporate the acid solution to a few cubic centimetres, add excess of soda solution, boil, and filter off any manganese precipitated, redissolve it in hydrochloric acid, and add the liquid to the acetic acid solution. Pass sulphuretted hydrogen through the alkaline filtrate and treat the zinc sulphide as in No. XIII. Part II. p. 106.

The filter containing the nickel, cobalt, and copper is dried and incinerated, dissolved in a few drops of aqua regia, treated with sulphuretted hydrogen, filtered if necessary, and the nickel and cobalt separated as in No. XVI. Part IV.

To the solution containing the manganese, add sodium carbonate in slight excess, boil, filter, wash the manganous carbonate, and ignite it until the weight is constant. It consists of mangano-manganic oxide. Care must be taken to remove the precipitate as completely as possible from the filter before it is incinerated.

Determination of the Phosphorus, Copper, Arsenic, and Antimony.—Weigh out about 10 grams of the iron in a state of fine powder into a capacious flask, place a small funnel in the neck, and pour strong nitric acid in small quantities at a time over the metal. Warm the liquid, and when all visible action is at an end, heat the residue with a fresh portion of nitric acid. Mix the solutions, add hydrochloric acid and evaporate to complete dryness: again add hydrochloric acid and evaporate a second time to dryness. Dissolve in water, and saturate the liquid with sulphuretted hydrogen. After the precipitate has completely settled, filter into a large flask, dry the precipitate, and digest it with carbon disulphide to remove the sulphur. Separate the copper, arsenic, and antimony in the black residue which remains, according to No. XVI. Part IV.

Pass a current of carbonic acid through the filtrate contained in the flask to dissipate the dissolved sulphuretted hydrogen, add a·small quantity of ferric chloride solution, nearly neutralise with sodium carbonate, add a little barium carbonate, and cork the flask ; after standing about twenty-four hours, the precipitate will contain the whole of the phosphoric acid ; filter it off, wash, dissolve in hydrochloric acid, remove the barium by the addition of sulphuric acid, filter, concentrate the filtrate, add molybdic acid solution, and proceed as in No. XIX. Part IV. p. 218.

The residue unattacked by aqua regia not unfrequently contains phosphide of iron. Fuse it, therefore, with sodium carbonate and nitre, extract with water, and test the solution for phosphoric acid.

Determination of Admixed Slag in Cast Iron.

By treating a piece of the metal by Weyl's method (p. 231), with very dilute hydrochloric acid, any admixed slag is not decomposed, and remains with the graphite, &c. The residue is collected on a filter, washed, and ignited in a platinum crucible until the carbon is consumed. The ignited mass is boiled with solution of sodium carbonate to dissolve the free silica, the insoluble matter ignited first in a stream of hydrogen, and then in dry chlorine (free from air), and treated with dilute hydrochloric acid, and again with boiling sodium carbonate solution. Filter off the insoluble matter, dry it, and weigh it as slag.

The following analysis by Fresenius of 'Spiegel-eisen,' produced from the spathic ore of Stahlberg, near Müsen, was made by the foregoing method :—

Iron .	.	. 82·860	Magnesium	.	0·045	Nitrogen	.	. 0·014
Manganese	.	10·707	Calcium .	.	0·091	Silicon	.	. 0·997
Nickel	.	. 0·016	Potassium	.	0·063	Carbon	.	. 4·323
Cobalt	.	trace	Arsenic .	.	0·007	Slag .	.	. 0·665
Copper	.	. 0·066	Antimony	.	0·004			
Aluminium	.	0·077	Phosphorus	.	0·059			100·014
Titanium .	.	0·006	Sulphur .	.	0·014			

XXII. Iron Slags.

The slag produced in the manufacture of cast iron may be regarded as a double silicate of lime and alumina, in which a portion of the lime is replaced by ferrous and manganous oxides, magnesia, and alkalies. Phosphoric acid and calcium sulphide are also usually present. The average composition of the blast-furnace slag is represented by the following analysis :—

Silica	.	. 41·85	Lime	.	. 30·99	Sulphur .	.	0·92
Alumina .	.	14·73	Magnesia.	.	4·76	Phosphoric acid		0·15
Ferrous oxide .		2·63	Potash	.	1·90			100·32
Manganous oxide	1·24	Calcium .	.	1·15			

Occasionally, however, either from carelessness or from defective working of the furnaces, the amount of iron in the slag is considerably augmented. Some slags are completely decomposed by treatment with hydrochloric acid : others are only partially acted upon. All of them yield more or less sulphuretted hydrogen on heating with hydrochloric acid. The amount thus evolved may be determined by one of the methods described on p. 239.

The methods of analysis described in Nos. XI. and XII. Part II. are generally applicable to slags which resist the action of hydrochloric acid.

A very convenient method of decomposing slags is to heat them in a finely-divided state in a sealed tube for two hours for 200° in an air bath, with a mixture of three parts by weight of strong sulphuric acid and one of water, or with hydrochloric acid containing 25 per cent. HCl. About one gram of the powder is introduced into a strong tube of Bohemian glass rounded at one end ; the other end is thickened before the blow-pipe. and drawn out. Add the acid, and carefully seal the tube. When the action is at an end, open the tube after cooling, and rinse out its contents into a porcelain dish, and evaporate to dryness in the ordinary way to render the silica insoluble. The remainder of the analysis is conducted after Nos. XI. and XII. Part II. This process is often applicable to the analysis of natural silicates : the determination of mixtures of ferric and ferrous oxides may also be conveniently made by this method.

Tap-cinder may be analysed by the above method, or according to the processes given in Nos. XX. XXI. Part IV.

XXIII. Assay of Zinc-Ores.

From 0·5 gram to 1 gram (according to its supposed richness) of the finely-powdered ore is dissolved in aqua regia, the solution is evaporated to dryness, the residue heated with water, filtered, and mixed with ammonium carbonate and ammonia solutions. The liquid is gently heated on a sand-bath for half an hour, filtered, and the insoluble matter washed with ammonium acetate solution. The filtrate containing the zinc is mixed with an excess of sodium or ammonium sulphide, and after standing for a short time the
ipernatant liquid is poured on to a large ·filter ; the pre-

cipitate is washed first by decantation and afterwards on the
filter with warm water containing ammonia, until the filtrate
no longer discolours an alkaline solution of lead acetate.
The filter is pierced, and the zinc sulphide is carefully
washed down into a 500 c.c. flask containing an excess of
ferric chloride solution and some free hydrochloric acid.
The flask is well closed and set aside in a warm place for a
short time. The mixture should occasionally be agitated to
accelerate the reaction. The zinc sulphide in contact with
the ferric chloride and free hydrochloric acid is converted
into zinc chloride, sulphur is separated, and the iron is reduced to the state of ferrous chloride.

$$ZnS + Fe_2Cl_6 = ZnCl_2 + 2FeCl_2 + S.$$

The solution (which should have a yellow colour, indicating an excess of ferric chloride, and be free from any smell
of sulphuretted hydrogen) is diluted to the containing-mark,
well shaken, and an aliquot portion of the liquid titrated
with acid-chromate of potassium solution. If the liquid is
quite cold and dilute, the free sulphur exercises no influence
upon the result.

The alkaline solution of zinc may also be titrated with a
solution of sodium sulphide of known strength. This method
is largely used in many zinc works on the Continent.
Saturate a solution of soda with sulphuretted hydrogen, and
add a second quantity of the alkaline liquid until the smell
of the gas is no longer perceptible. Dissolve 10 grams of
pure zinc in dilute sulphuric acid (taking care not to use a
very great excess of acid), and dilute the solution to 1 litre.
Also dissolve a few crystals of sodium tartrate in water, add
a small quantity of caustic soda and lead acetate, and heat
the mixture until the liquid is clear.

Transfer 25 c.c. of the standard zinc solution to a beaker;
add a mixture of ammonium carbonate and ammonia sufficient to redissolve the precipitate first formed, and spread a
few drops of the alkaline lead solution on a piece of filter-
paper placed on a porcelain slab or plate, and run in the
solution of sodium sulphide until a drop of the liquid with-

drawn by a glass rod and brought into contact with the lead acetate forms a black ring at the point of contact. If necessary, the solution of the sodium sulphide is diluted to a convenient strength for titration, and its exact value again determined in the above manner on quantities of 25 c.c. or 50 c.c. of the standard zinc solution.

The amount of zinc in the ammoniacal liquid obtained by treating the ore is then determined in exactly the same manner by the sodium sulphide solution. The strength of the solution of sodium sulphide must be re-determined before each series of experiments, as it experiences alteration by exposure to the air.

Many zinc-blendes contain notable quantities of copper, which by combining with the sodium sulphide increases the amount of zinc apparently present. The safest plan in such a case is to remove the copper by sulphuretted hydrogen in an acid solution ; filter, evaporate the filtrate with nitric acid, dilute, add ammonia, and proceed as directed.

In the case of ores which contain alumina and manganese, zinc is more accurately estimated by a standard solution of potassium ferrocyanide. The solution of the ore prepared as above is acidified with hydrochloric acid, and the ferrocyanide solution (previously standardised by means of pure zinc) is added until a drop withdrawn from the liquid gives a brown colour with solution of uranium nitrate placed on a porcelain slab. (Comp. pp. 333, 334.)

XXIV. Assay of Tin-Ores.

The amount of tin in its ores may be easily estimated by fusing with potassium cyanide : the process, although not absolutely correct—especially in presence of lead or copper —yields results of sufficient accuracy for technical purposes. About 6 or 8 grams of the finely-powdered ore is placed in a smooth porcelain mortar, and intimately mixed with five times its weight of commercial potassium cyanide. The mixture is projected into a small clay crucible, in the bottom of which a quantity of powdered potassium cyanide has been previously placed, sufficient to form a layer of 1 or 2 centimetres in depth, and the mortar is rinsed with a fresh quantity of cyanide, which is poured on the top of the mixture. It is advisable by way of control to prepare two such crucibles, and to take the higher result as the true one. The crucibles are heated to a moderate red heat, and the cyanide is kept in fusion for 10 or 15 minutes : they are removed from the fire and gently tapped, to promote the formation of a single button of the reduced metal. After cooling they are broken, and the buttons of metal are extracted and weighed after the adhering flux has been removed. The saline mass should be triturated with water in order to be certain that the reduction has been effectual and that the whole of the metal has been collected in one piece. The silica in the ore unites with the alkaline carbonate always contained in the commercial cyanide. If the ore contains any considerable portion of lead or copper, these must first be removed by digestion with strong hydrochloric acid.

XXV. Separation of Tin from Tungsten.

Commercial stannates of soda are frequently mixed with sodium tungstates. The best method of determining the proportion of the two acids is to fuse the mixture with potassium cyanide, when the tin only is reduced to the metallic state. About 2 grams of the powdered salt is mixed with four times its weight of fused and powdered potassium cyanide, and the mixture heated in a porcelain crucible. By treating the fused mass with water the alkaline tungstate dissolves, together with the excess of the cyanide : the reduced metal is washed, and converted into oxide by treatment with nitric acid. The filtrate is heated with nitric acid to decompose the potassium cyanide, evaporated nearly to dryness, the residue dissolved in alkali, acidified with nitric acid, and the tungstic oxide precipitated by means of mercurous nitrate. The mercurous tungstate is washed with water containing a little mercurous nitrate, dried, and converted into tungstic oxide by ignition in contact with air.

XXVI. WOLFRAM.

This mineral is a tungstate of iron and manganese ($FeMn''$) WO_4, in which the proportion of the two metals is variable. It occurs associated with tin-ore, tungstate of calcium, galena, &c., in Cornwall, Cumberland, in the Hebrides, France, Bohemia, and North America.

The finely-divided mineral is heated with aqua regia until it is completely decomposed. The solution is evaporated to complete dryness over the water-bath, water added, and the manganese and ferric chlorides filtered off. The residual tungstic acid is washed with alcohol, dissolved in ammonia, the solution filtered from any residue, evaporated to dryness in a capacious porcelain crucible, gently heated to expel ammonia, and ignited in contact with air. The tungstic oxide is then weighed : it should have a pure yellow colour, free from any greenish tint.

Test the residue for niobic acid* by heating it in the Bunsen flame with microcosmic salt : the oxide of niobium forms a colourless bead in the outer flame, but a violet-coloured bead inclining to blue in the inner flame if the fused salt is saturated with the oxide. By adding a trace of ferrous sulphate the colour is changed to blood-red. As very similar reactions are afforded by tungstic oxide, care must be taken to ensure that the whole of this substance is removed by ammonia before the test is made. Another portion may be heated with a bead of sodium carbonate, by which it is dissolved, forming whilst hot a transparent mass, which becomes turbid on cooling ; if whilst still hot it is moistened with a drop of tin chloride, and heated in the lower reducing flame, it gives a grey mass, which dissolves in hydrochloric acid, producing a light amethystine tint.

The alcoholic filtrate is evaporated to dryness, the residue dissolved in water, and the iron and manganese separated as in No. XIX. Part IV. p. 220.

* Columbite, a niobate of iron and manganese, not unfrequently occurs associated with wolfram.

XXVII. Scheelite (CaWO$_4$).

This mineral is readily decomposed by strong nitric acid. The solution is evaporated nearly to dryness, alcohol added, and the liquid filtered. Tungstic oxide is left undissolved : calcium nitrate goes into solution. The oxide is treated in the manner above described : the solution is evaporated nearly to dryness, water added, and the lime estimated in the usual manner.

XXVIII. Galena.

This substance is essentially lead sulphide, but it is almost invariably mixed with more or less iron, copper, silver, antimony, and zinc, and silicious matter (gangue).

Determination of the Sulphur.—The ore is reduced to the finest powder and dried at 100°. About 1 gram of the substance is weighed out into a large porcelain crucible, gently heated with potash solution (free from sulphate) for an hour, and a slow current of chlorine conducted into the liquid. The galena by this treatment is decomposed ; the sulphur is oxidised to sulphuric acid, which combines with the potash, and the lead is converted into binoxide. The liquid is filtered, acidified with hydrochloric acid, and the sulphur precipitated as barium sulphate.

Determination of the Lead, Iron, Zinc, &c.—About 1·5 to 2 grams of the ore are oxidised with red fuming nitric acid (B.P. 86°) in a flask, in the mouth of which is placed a small funnel. The sulphur is thus completely oxidised, and the galena converted into lead sulphate. Evaporate off the excess of acid, and add about 5 cubic centimetres of moderately-strong sulphuric acid, and evaporate nearly to dryness. Add about 20 cubic centimetres of water, filter, wash the residue with water containing sulphuric acid, and remove the sulphuric acid by washing with alcohol, otherwise the paper will fall to pieces on being dried. Do not mix the alcoholic washings with the acid filtrate. The operation of washing must be done without delay, and with as little water as possible, otherwise a perceptible amount of lead sulphate will be dissolved. The residue in the filter is dried and ignited in a weighed porcelain crucible ; care must be taken to remove as much of the lead sulphate as possible from the paper before incineration. The ash may be moistened with one drop of nitric acid and then one drop of sulphuric acid, and the whole carefully dried, ignited, and weighed. The substance consists of lead sulphate mixed with sand, silica (gangue). When weighed, it is carefully transferred to a small beaker and heated with hydrochloric acid, which dissolves the lead sulphate, leaving the silicious matter unchanged. Allow the liquid to become clear by standing, and pour it through a small filter, again digest with hydrochloric acid and again filter, repeating this treatment three or four times until the filtrate is no longer blackened by sulphuretted hydrogen water. Wash the residue on to the filter with hot water, dry and weigh, and subtract the weight, minus the filter ash, from the original weighing : the difference gives the amount of lead sulphate.

Pass sulphuretted hydrogen through the filtrate, and determine the copper and antimony as in No. XVI. Part IV. Iron and zinc are precipitated from the filtrate, after treatment with sulphuretted hydrogen, by means of ammonium sulphide : they may be separated as in No. XVI. Part IV.

Determination of Silver in Galena.—Galena rarely contains as much as 0·5 per cent. of silver, but this metal can be profitably extracted, even when it does not exceed one-twentieth of this amount in the ore. The exact determination of the small quantities of silver almost universally present in galena becomes, therefore, a matter of importance. When an argentiferous galena is smelted, the whole of the silver is found in the reduced metal. About 50 or 60 grams of the finely-divided galena are mixed with twice their weight of sodium carbonate and 20 grams of nitre, and placed in a clay crucible. A layer of well-dried common salt (about 8 or 10 millimetres deep) is placed over the mixture, and the crucible is heated to bright redness. It is allowed to cool, broken, and the button of reduced lead extracted. This is flattened on an anvil and freed from slag, &c., by rubbing and washing with water. The following equation represents the reaction:

$$4Na_2CO_3 + 7PbS = 4Pb + 3(PbSNa_2S) + Na_2SO_4 + 4CO_2.$$

The object of the nitre is to decompose the double sulphide of lead and sodium : the lead is separated, and the sodium sulphide oxidised to sulphate. The button of lead (which should weigh from 35 to 40 grams, if the operation has been properly conducted) is slowly dissolved in pure dilute nitric acid until about 5 or 10 grams of the metal remain : this is withdrawn from the solution. It contains the whole of the silver. It is dissolved in dilute nitric acid, and the solution is diluted with a large quantity of water. A few drops of highly-dilute hydrochloric acid are added, and the liquid is allowed to stand until the silver chloride has completely settled : this is filtered off, washed repeatedly with hot water, and weighed. (Compare also No. XXXIV. p. 279.

The object of only partially dissolving the button is to avoid the presence of an undue amount of lead in solution, since the nitrate of this metal dissolves silver chloride to a perceptible extent. The results obtained by this method, if properly conducted, are more exact than those given by cupellation.

A very ready method of assaying galena for technical purposes consists in decomposing the ore by means of zinc and hydrochloric acid. About 2 grams of the finely-powdered sulphide are weighed out into a tall beaker and covered with a piece of pure zinc about an inch in diameter and a quarter of an inch thick.* Pour into the beaker about 120 cubic centimetres of dilute hydrochloric acid (1 part acid to 4 of

* Obtained by dropping the molten metal upon a smooth surface of wood or metal.

water), cover the beaker with a watch-glass, and gently heat (to 40° or 50°) for 15 or 20 minutes, occasionally stirring the liquid. When the evolution of sulphuretted hydrogen ceases, and the liquid becomes clear, the decomposition is complete. The supernatant liquid is poured on to a filter, in which a small piece of zinc is placed, and the zinc and lead in the beaker washed with hot water by decantation until the filtrate has no longer an acid reaction. The lead is transferred to a weighed porcelain crucible, any portions adhering to the zinc being rubbed off by a glass rod. The small particles on the filter are washed into a porcelain basin and added to the crucible. The water in the crucible is poured away and the lead gently dried in a current of coal gas to prevent it from oxidising. If gangue is mixed with the reduced lead, its amount may be determined by dissolving the metal in dilute nitric acid, and washing, drying, and weighing the insoluble residue.

XXIX. REFINED LEAD.

Recent improvements in refining, and the introduction of such improved methods of desilverisation as Pattinson's crystallisation process or Parkes' zinc process, have so far perfected the process of manufacturing softened lead that this article seldom contains less than 99·9 per cent. of the pure metal. Pure as such lead may appear, the presence in it of minute traces of iron, copper, &c., yet exercises a very important influence in its application to the manufacture of vitriol chambers, evaporating pans, and to the preparation of white lead, flint glass, &c.

The following substances are found in refined lead : silver, copper, bismuth, cadmium, zinc, iron, nickel, and antimony. Cobalt, arsenic, and manganese are seldom present in estimable quantity. These metals are derived partly from the ores and partly from the employment of Parkes' process, the impurities being introduced in the zinc used.

*The Method of Analysis.**—The lead to be analysed is cut up into large pieces, and the surface of each piece is scraped with a clean bright knife until it is perfectly bright and free from any apparent impurity. Weigh off two portions of 200 grams each of the lead into flasks of 1,500 cubic centimetres capacity, and add to one portion about 500 cubic centimetres of *pure* nitric acid of sp. gr. 1·2, and so much water that no lead nitrate separates out. The action of the acid may be promoted by a gentle heat ; care must be taken not to employ a greater excess of nitric acid than that given. The solution is allowed to stand from 12 to 24 hours. Since 200 grams lead give 310 grams of nitrate, and 1 part of lead nitrate requires about 2 parts of water for solution, there is no possibility of lead nitrate separating out from the solution if it measures about 1 litre. If a crystallisation occurs, it is a sign that too great an amount of nitric acid has been used, lead nitrate being far more insoluble in dilute nitric acid than in water.

To the other 200 grams add nitric acid of the above strength in small portions at a time, always keeping the metal in excess, and heat the liquid until only about 5 or 10 grams of lead remain undissolved, and the solution commences to turn yellow in consequence of the formation of lead nitrite. In the residual metal the whole of the silver is concentrated. It is withdrawn from the liquid, dissolved in nitric acid, the solution diluted with a large quantity of water, and a few drops of a solution of lead chloride added, or 1 cubic centimetre of hydrochloric acid of 1·12 sp. gr., previously diluted with 50 cubic centimetres of water. This quantity of acid is more than sufficient to precipitate all the silver without throwing any lead chloride out of solution. Set the beaker aside for two or three days. Draw off the clear liquid by means of a syphon, and collect the precipitate on a small filter, wash thoroughly with boiling water, dry it,

* Fresenius, Zeits. für anal. Chem. vol. viii. 1869.

and incinerate in a small weighed porcelain crucible. If the amount of silver chloride is so considerable that there is a possibility of its being incompletely reduced by the combustion of the filter-paper, the residue must be heated for a few minutes in a stream of hydrogen before weighing. The amount left, after subtracting the filter ash, gives the quantity of silver in the 200 grams of lead. The refined metal seldom contains more than 0·0015 per cent. of silver. (Mean of 12 specimens, 0·0013 per cent.)

The solution of the other portion of 200 grams is used for the estimation of the remaining impurities. As a rule it remains perfectly clear even after standing. Occasionally, however, in the case of lead rich in antimony, a more or less considerable precipitate forms after a time. The precipitate is filtered off and set aside ; the mode of examining it will be given hereafter. Bring the clear solution or filtrate into a 2-litre flask, add 115 grams (about 62 or 63 cubic centimetres) pure and concentrated sulphuric acid, shake, allow to cool, and fill up to the mark ; again shake the liquid so as to mix it thoroughly, and allow the precipitate to settle. The amount of sulphuric acid to be added is so calculated that about 10 or 12 grams are in excess. As soon as the lead sulphate has completely settled, the clear liquid is drawn off by means of a syphon previously filled with the liquid. In this way rather more than 1,750 cubic centimetres may be drawn off. Accurately measure off 1,750 cubic centimetres of the solution, and evaporate it in a porcelain basin, in a draught-place free from dust, until sulphuric acid fumes make their appearance, indicating that the nitric acid has been expelled. Allow the liquid to cool, add about 60 cubic centimetres of water, and filter off the small quantity of lead sulphate which separates out, and wash the precipitate with a little water.

The slight precipitate of lead sulphate frequently retains small quantities of antimony. It is therefore dissolved in hydrochloric acid, and the solution diluted with sulphuretted hydrogen water, the liquid warmed, and sulphuretted hydro-

gen passed through it. The precipitate is allowed to subside, filtered, washed, the filter-paper spread out in a
porcelain basin, and heated with a solution of pure ammonium or potassium sulphide, to which a small quantity of
pure sulphur has been added. Filter the solution, wash,
acidify the filtrate with hydrochloric acid, and allow the
precipitate to settle at a gentle heat.

The solution filtered from the lead sulphate, and diluted
to 200 cubic centimetres, is heated to about 70°, and treated
with sulphuretted hydrogen, allowed to stand 12 hours at
a gentle heat, and filtered through a small filter. The
washed precipitate is heated with potassium sulphide solution (containing sulphur) in the manner above described.
The filtrate is acidified with hydrochloric acid and allowed
to stand until the precipitate has completely subsided.

The filter and residue insoluble in potassium sulphide are
heated in the dish nearly to boiling with dilute nitric acid
(1 part acid 1·2 sp. gr. and 2 of water). When the precipitate is dissolved, filter, wash the paper slightly, dry, and
incinerate, and add the ashes to the nitric acid solution ;
add 2 cubic centimetres of dilute sulphuric acid, and evaporate until the nitric acid is expelled ; dilute with a little
water, and filter off the small quantity of lead sulphate which
separates out. Nearly neutralise the filtrate with pure caustic
potash, add sodium carbonate, and a small quantity of
potassium cyanide solution free from potassium sulphide,
and heat gently. If a precipitate is formed, it is filtered off,
washed, and dissolved in dilute nitric acid, and the bismuth
precipitated by ammonium carbonate, and weighed as oxide.
To the filtrate a little more potassium cyanide is added,
together with a few drops of potassium sulphide. The
precipitate which ensues contains the cadmium and silver.
It is filtered off, dissolved in dilute nitric acid, the silver
precipitated by hydrochloric acid, the filtrate evaporated
nearly to dryness, and a few drops of sodium carbonate
solution added. The cadmium precipitate is filtered off,

dried, ignited, and weighed as oxide. The reduction and volatilisation of the cadmium may be prevented by moistening the filter with ammonium nitrate. The filtrate from the mixed silver and cadmium sulphides is mixed with a small quantity of sulphuric and nitric acids, and evaporated nearly to dryness ; a few drops of hydrochloric acid are added, and the solution heated until the last traces of hydrocyanic acid have disappeared ; the solution is filtered, if necessary, and the copper precipitated and weighed as sulphide.

When cadmium is absent, the separation of the copper and bismuth may be effected by means of ammonia and ammonium carbonate. In this case it must not be forgotten that the silver is to be removed by the addition of hydrochloric acid before the copper is precipitated as sulphide.

The precipitates obtained by acidifying the sulphide of ammonium solution are filtered off, washed, dried, and repeatedly treated with carbon disulphide to remove the sulphur. The little filter and the residue are then together warmed with a few drops of red fuming nitric acid, the porcelain basin being covered with a watch-glass ; the solution is heated to expel the excess of nitric acid, sodium carbonate added in slight excess, and then a small quantity of sodium nitrate. The solution is evaporated to dryness, and carefully heated until the residue melts and the mass becomes white. When cold the fused mass is transferred to a small mortar and carefully broken up by rubbing it in a little cold water. The solution is filtered and the residue is washed with water containing alcohol. The sodium antimoniate is dissolved in hydrochloric acid, to which a little tartaric acid has been added, and the solution treated with sulphuretted hydrogen, and set aside for a few hours.

The soluble portion of the fused mass which contains all the arsenic, together with a little antimony, is evaporated to expel the alcohol, an excess of sulphuric acid is added, and the solution again evaporated to expel the nitric acid, water added, the liquid heated to 70° C., and treated with sulphuretted hydrogen. When the precipitate has settled, it is filtered through a small filter and washed with water. It is then treated on the filter with a cold concentrated solution of ammonium carbonate, the filtrate being repeatedly poured back on to the filter, to obviate the use of a large excess of ammonium carbonate. The arsenic sulphide is dissolved: the antimony sulphide mixed with a little sulphur remains undissolved. This residue is dissolved in a little strong hydrochloric acid diluted, and treated with sulphuretted hydrogen, and the precipitated sulphide added to the main quantity obtained from the sodium antimoniate. The antimony sulphide is best filtered through a small tube in the bottom of which a little asbestos has been placed. The tube containing the asbestos is gently heated over the direct flame and weighed. When the whole of the precipitate has been transferred to the little tube, it is warmed to expel the greater portion of the water, and then gently heated in a stream of dried carbon dioxide until the antimony sulphide becomes black. The tube is allowed to cool in a current of the gas, the carbon dioxide displaced by atmospheric air, and the tube and antimony sulphide again weighed.

The solution of arsenic sulphide and ammonium carbonate is acidified with hydrochloric acid, and the turbid solution treated with a little sulphuretted hydrogen water, filtered, and the arsenic sulphide filtered through a weighed tube containing asbestos, and heated in the manner prescribed in the case of the antimony sulphide.

The filtrate and washings from the original precipitate by sulphuretted hydrogen are concentrated, poured into a flask, rendered alkaline by ammonia, and mixed with ammonium sulphide. The flask, which must be at least half full of liquid, is well corked and allowed to stand 24 hours. When the slight precipitate has subsided the liquid is filtered, the filtrate acidified with acetic acid, and evaporated at a gentle heat to facilitate the separation of a small quantity of nickel sulphide. This, mixed with sulphur, is filtered off, slightly washed, and dried and incinerated.

The precipitate formed by the ammonium sulphide is treated on the filter with a mixture of 1 part hydrochloric acid (sp. gr. 1·12) and 6 parts sulphuretted hydrogen water, the filtrate being repeatedly poured back on to the filter. The sulphides of iron and zinc are dissolved: the nickel and cobalt sulphides remain on the filter. The filter is dried, incinerated, and mixed with the ash of the filter containing the nickel sulphide derived from the sulphide of ammonium solution. The mixed ashes are treated with aqua regia, the solution concentrated, a little water added, filtered, rendered alkaline by ammonia, a few drops of ammonium carbonate added, filtered into a platinum basin and treated with a few drops of potash solution until no more ammonia is evolved. Filter off the slight flocculent precipitate, wash, dry, ignite, and weigh, and test the precipitate (which generally consists mainly of nickel oxide) for traces of cobalt with the blowpipe. The solution containing the iron and zinc, to which a few drops of nitric acid are added, is concentrated by evaporation, and rendered alkaline by ammonia, the ferric oxide filtered off, again dissolved in a few drops of hydrochloric acid, and again precipitated by ammonia, filtered, washed, dried, and weighed. By way of control the weighed precipitate may be fused with potassium-hydrogen sulphate, dissolved in water, and reduced with zinc, and the iron estimated volumetrically by a dilute permanganate solution.

The filtrate from the ferric hydrate is mixed with a little ammonium sulphide, and allowed to stand at least for 24 hours at a gentle heat. If anything separates out it is filtered off, washed, and digested on the filter with dilute acetic acid in order to separate any admixed manganese sulphide. The residue on the filter is dried, and weighed as zinc sulphide. The acetic acid solution is concentrated to a few cubic centimetres, and mixed with excess of caustic potash to precipitate the manganese.

Before we can proceed to express the results centesimally

it is necessary to determine the quantity of lead corresponding to the 1,750 cubic centimetres of solution taken for analysis. This can only be estimated when we know the volume occupied by the lead sulphate obtained from the 200 grams of metal, when suspended in water. By repeated experiments it has been found that it occupies the space of 44·99 grams, or in round numbers, 45 grams of water at 16° C. The 2-litre flask, when filled to the mark, holds therefore 1,955 cubic centimetres solution, and 45 cubic centimetres lead sulphate. But the 1,955 cubic centimetres of solution were equivalent to 200 grams lead; therefore the 1,750 cubic centimetres would be equal to 179·03 grams, or in round numbers, 179 grams of the original lead.

The solution of the lead, when containing unusually large quantities of antimony, not unfrequently forms, on standing, a white precipitate of antimony oxide and antimoniate of antimony, which occasionally retains arsenic. This precipitate is filtered off, washed, and the arsenic and antimony separated as above. In calculating the result, it must not be forgotten that this precipitate is obtained from 200 grams of the metal, whilst the remaining portion in solution is assumed to be derived from 179 grams of lead.

The determination of the minute quantity of sulphur contained in lead may be effected by heating the metal in chlorine gas, when the sulphur is converted into the chloride, which, when led into water, is decomposed, with the formation of sulphuric acid : this may be precipitated in the usual way, and weighed as barium sulphate. The best method of carrying out this process is to heat about 100 grams of the lead in the form of a thick rod about 1 centimetre in diameter in a combustion-tube about 1 metre long. In the middle the tube is narrowed, and the end is drawn out and bent downwards, and dips into a small three-bulbed U-tube containing water. The lead is placed in the anterior portion of the tube ; the other serves to collect the lead chloride, which flows over the little bridge as fast as it is produced, leaving the metal exposed to the further action of the gas. The combustion-tube is connected with a small tube containing ignited fragments of charcoal, over which the chlorine passes before it comes in contact with the heated lead. The charcoal must be kept at a red heat throughout the operation : it serves to free the chlorine from any trace of accompanying oxygen, which might oxidise the lead sulphide to sulphate. Vulcanised stoppers, on account of the sulphur they contain, cannot be used to connect together the several pieces of the apparatus. Ordinary corks must therefore be employed. When the entire apparatus is filled with chlorine the lead is gradually melted, care being taken to place the combustion-tube in such a position that the metal does not come in contact with the cork ; it should flow against the bridge, but not above it.

When the heat is properly regulated the lead burns slowly to lead chloride, which flows over and collects in the empty portion of the tube. When care is taken to regulate the stream of chlorine and not to overheat the chloride, but little of this substance passes over into the U-tube. The solution is washed out of the condensing tube into a small beaker, heated, and the sulphur precipitated by the addition of a few drops of barium chloride.* The following analysis of three specimens of soft lead, executed by the above methods, will give some idea of the nature and amount of its impurities : the results are represented centesimally :—

Silver	.	. 0·00200	0·00062	0·00385
Copper	.	. ·00228	·00031	·00190
Cadmium	.	. trace	·00010	—
Bismuth	.	. ·00040	·00010	·00553
Antimony	.	. ·00173	·00186	·02639
Iron	.	. ·00035	·00012	·00129
Zinc	.	. ·00014	·00023	—
Sulphur	.	. ·00076	·00008	—

* Bannow and Krämer, 'Ber. Deutschen Chem. Gesells.' July 1871.

XXX. WHITE LEAD.

This substance is a compound of lead carbonate and hydrate in variable proportions. In general the relation between the hydrate and carbonate may be represented by the formula $2PbCO_3 + PbH_2O_2$, although specimens of the composition $3PbCO_3 + PbH_2O_2$ and $5PbCO_3 + PbH_2O_2$ are occasionally obtained (Mulder, Phillips). As found in commerce it is frequently mixed with barium sulphate (heavy spar), barium carbonate (witherite), calcium carbonate, and zinc oxide. These bodies cannot always be regarded as adulterants ; the heavy spar, for example, serves to protect the lead from the rapid action of sulphuretted hydrogen, and unless present in large excess does not interfere with the opacity or *body* of the pigment. But these substances are not unfrequently added in undue quantity ; and perfectly pure white lead is now comparatively rare as an article of commerce.

Determination of the Carbon Dioxide.—From 1 to 2 grams of the substance are weighed out into the flask A, fig. 31, and decomposed by moderately-dilute nitric acid. The operation is carried out exactly in the manner described on p. 86.

On the termination of the experiment the liquid in the flask A is filtered if necessary, and the residue washed, dried, and weighed : it may consist of the sulphates of barium or calcium. The weighed residue is boiled in a platinum or porcelain basin with solution of pure sodium carbonate for an hour or so, care of course being taken to replenish the dish with water from time to time. Any calcium sulphate present is completely decomposed, and, on pouring the liquid through a small filter, the sulphate in solution may be detected by the addition of barium chloride. The calcium carbonate formed may be dissolved out by dilute hydrochloric acid, and the lime precipitated by means of ammonium oxalate. The washed precipitate is rendered caustic by ignition and weighed.

The residual barium sulphate may also be weighed, by way of control.

The nitric acid filtrate containing the lead, &c., is evaporated nearly to dryness to expel the excess of the acid, diluted with water, and the liquid saturated with sulphuretted hydrogen. The lead is completely separated; it is filtered off, and converted into sulphate by oxidation with nitric acid. The weighed precipitate should then be heated with a dilute solution of sodium thiosulphate, which dissolves the lead sulphate and leaves unattacked any barium sulphate which may be present. To the filtrate from the lead sulphide add ammonia and ammonium sulphide; wash the precipitated zinc sulphide, dissolve it in dilute hydrochloric acid, and re-precipitate as carbonate, and convert into oxide by ignition. The filtrate from the zinc sulphide contains the lime and baryta : these are separated as in No. VII. Part II. If baryta is found no calcium sulphate can be present. If it is desired to determine the water directly, this can be effected by means of the apparatus shown in fig. 53.

Arrangement of the Results.—Calculate the baryta found in the last filtrate to barium carbonate, and the lime to sulphate or carbonate according to circumstances. The residual amount of carbon dioxide is converted into lead carbonate, and the remainder of the lead to lead oxide. The water, zinc oxide, and barium sulphate are set down as such in the statement of the analysis.

Zinc white may be also analysed by the foregoing method. In addition to the adulterants mentioned above, kaolin is not unfrequently met with : this is left undissolved on treating the pigment with dilute acid.

XXXI.   Chrome Iron-Ore.

This mineral occurs massive in various parts of the world, particularly in Norway, Siberia, Asia Minor, Silesia, and North America ; it consists essentially of a compound of ferrous oxide and chromium sesquioxide. It belongs to the spinelle group of minerals, and is isomorphous with magnetic oxide of iron. Its formula, $FeOCr_2O_3$, requires 67·7 per cent. of chromium sesquioxide. Usually, however, the chromium is replaced by aluminium, and the iron by magnesium to a considerable extent, and the ore is of very good quality when it contains 50 per cent. of chromium sesquioxide.

Determination of the Chromium Sesquioxide.—Grind a few grams of the carefully-sampled mineral in an agate mortar, and pass the powder through a fine muslin sieve. When you have collected about two grams of the fine dust, again grind it in portions of a decigram at a time in the agate mortar until every feeling of grittiness has disappeared, and the ore cakes in an impalpable powder round the pestle. Too much care cannot be given to the grinding : the success of the analysis entirely depends upon the ore being in the finest possible state of division. Weigh out about 0·5 gram of the powder into a platinum crucible of about 50 cubic centimetres capacity, and place over it 12 times its weight of recently-fused potassium-hydrogen sulphate, and gently heat the crucible so as merely to melt the sulphate. Keep it melted at a gentle heat for 15 or 20 minutes, and gradually increase the temperature until the bottom of the crucible becomes red hot. Care must be taken that the fused mass does not rise above half way up the crucible. In a few minutes the mixture will fuse quietly and dense fumes of sulphuric acid will be evolved; the heat is now gradually increased until the crucible is at a bright red heat ; in about half an hour remove the lamp, and add about 6 parts of powdered sodium carbonate, again fuse the mixture, and, keeping the temperature for an hour at a red heat, add little by little the same quantity of nitre. The temperature of the crucible is now increased and kept at a full red heat for 20 minutes; it is allowed to cool, transferred to a porcelain basin, and the mass boiled out with water. The solution is filtered, and the residue washed with hot water until the filtrate comes through colourless. It is not necessary to transfer the whole of the insoluble matter to the filter. Quickly dry the filter and its contents, detach the ferric oxide and return it to the porcelain basin, burn the filter and add the ash, and digest the whole at a gentle heat with moderately-concentrated hydrochloric acid. If the fusion has been properly conducted the residue will dissolve ; any black insoluble matter left undissolved denotes that the grinding has not been done with sufficient care. This insoluble portion must be collected on a small filter, dried, and the filter, &c., folded up, thrown into the platinum crucible, ignited, and the residue again fused with potassium bisulphate, sodium carbonate, and nitre, and the fused mass again boiled out with water, filtered, and the filtrate added to the main solution. A few grams of ammonium nitrate are added to the total filtrate, and the liquid is evaporated nearly to dryness, water is added, and the solution is filtered from the alumina, silica, &c. The filtrate, which should be received in a porcelain basin,* is then made strongly acid with sulphurous acid, and boiled until the gas is nearly expelled, a slight excess of ammonia is added, and the solution is again boiled until it becomes colourless. Pour the liquid on to a filter, wash the precipitate by decantation with hot water, and by means of a feather transfer it to the filter, and wash *carefully* with hot water. After the sixth washing allow the filter and precipitate to drain thoroughly by keeping up the action of the pump for about ten minutes, remove the filter, fold it, and, without further drying, transfer it to a weighed platinum crucible and cautiously heat with the lid on. Gradually increase the heat, and as soon as the paper is charred, remove the lid, placing it at the edge of the crucible (see fig. 15), and ignite

---

* If the boiling with ammonia is conducted in glass vessels there is great probability that the precipitated chromic oxide will be contaminated with silica. The error from this cause may amount to 0·5 per cent. (Compare Souchay, Fres., 'Zeits.' iv. 66.)

strongly for 10 or 15 minutes. If the precipitate has been drained sufficiently by the action of the pump, there is no danger of the oxide being projected from the crucible on ignition. On treating the weighed precipitate with a few drops of water the liquid ought to remain colourless : a yellow colour indicates that the oxide has been imperfectly washed from alkaline salts.

Complete analysis of chrome iron ore (Dittmar's process).— Fuse together a mixture of equal weights of pure borax glass and sodium potassium carbonate in a platinum basin, and break up the fused mass when cold. 10 grams of this mixture are placed in a platinum crucible of about 50 c.c. capacity, fused, and allowed to cool. About 1 gram of the finely powdered ore is added, and the mass again fused, at first with the lid on : the crucible is now inclined, and the contents are stirred with a stout platinum wire. In about 15 minutes the whole of the ore should be dissolved. The lid is now placed in the position seen in fig. 15, the fused mass being occasionally stirred to promote its oxidation : this is usually complete in about half an hour. Allow to cool, and add about 5 grams of pure potassium carbonate, again fuse, and pour out the melted mass into a platinum basin, which should be quickly covered with a clock glass as the solidifying substance decrepitates. The portion adhering to the crucible is dissolved off by hot water in a porcelain basin, and the solidified mass in the dish is added to the solution. Add a few drops of alcohol to reduce any manganate which may be present, and digest on a water-bath until the whole is disintegrated and the alcohol expelled. Filter and wash the precipitate until the filtrate is colourless, and dissolve the precipitate in hot dilute sulphuric acid to ascertain if any ore is left undecomposed. If any be found it must be treated again with a small quantity of the fusion mixture, and the mass dissolved out with water as before. The filtrate is concentrated to about 200 c.c., and divided by weighing or measuring into two approximately equal portions. To the one portion is added a quantity of dilute sulphuric acid and a known weight (in excess) of pure ferrous sulphate, and the amount of un-oxidised iron determined by standard potassium bichromate solution as directed on p. 221.

$$Cr_2O_3 : 6Fe = 152\cdot2 : 336 = \cdot4529 : 1.$$

If only the amount of chromium is required, the determination may be repeated on the second portion. To determine the amount of silica and alumina, add ammonium nitrate to the solution, evaporate to dryness, add water, and filter off the alumina and silica.

The sulphuric acid solution is treated with sulphuretted hydrogen, filtered, and the filtrate concentrated to a small bulk, and poured into an excess of strong and fresh potash solution contained in a platinum basin, filtered, and the alumina and small quantity of silica precipitated by boiling with ammonium chloride : the precipitate is added to that obtained as above, and the substances are separated as in an ordinary silicate analysis. The separation of iron, lime, and magnesia is effected as described on p. 87.

XXXII. SMALTINE : COBALT-GLANCE.

Smaltine or tin-white cobalt is an arsenide of cobalt : cobalt-glance is essentially a sulpharseniate of cobalt. Both minerals frequently contain, in addition to cobalt, arsenic, and sulphur, variable quantities of nickel, iron, lead, bismuth, copper and gangue.

Determination of Sulphur.—See Copper Pyrites, p. 211.

Determination of Silica and the Metals.—Treat 2 grams of the finely-divided substance with strong nitric acid, and evaporate to dryness with sulphuric acid (hydrochloric acid is inadmissible, since a small quantity of the arsenic would be volatilised on heating). Moisten the dried mass with sulphuric acid and add water. On standing, the supernatant liquid ought to become quite clear : a turbidity indicates the presence of basic salts. Filter the liquid into a flask, wash the precipitate slightly with water containing sulphuric acid, remove the acid from the paper by alcohol, dry, and weigh; the insoluble matter consists of silica, and calcium and lead sulphates. Transfer the weighed precipitate to a small beaker and boil with dilute nitric acid: the sulphates are thus dissolved. Throw the residual silica on a filter, wash, dry, and weigh. The filtrate containing the lead and calcium is evaporated to dryness with hydrochloric acid, water added, the solution boiled, and the lead precipitated with sulphuretted hydrogen and converted into sulphate by treatment with strong nitric acid. The lead sulphate is weighed: its weight, *plus* that of the silica, subtracted from the original weight of the precipitate (SiO_2 + $PbSO_4$ + $CaSO_4$), gives the calcium sulphate.

To the filtrate from the insoluble residue ($SiO_2.PbSO_4$. $CaSO_4$) a strong solution of sulphurous acid is added in small quantities at a time, and the liquid boiled after each addition. The liquid is maintained at a temperature of 60° or 70° by surrounding it with warm water, and the metals precipitated by sulphuretted hydrogen. Allow the solution to stand for some time, so that the greater portion of the sulphuretted hydrogen may be removed by diffusion, and filter the liquid. Wash and drain the filter thoroughly by the action of the pump, spread it on a glass plate, and remove the arsenic sulphide by means of a thick platinum wire, transferring it to a large porcelain crucible. Fold the filter and replace it in the funnel, and dissolve the small quantity of adhering sulphide by a few drops of strong potash solution, and allow the liquid to fall into the porcelain crucible. A little more potash is added to the crucible until the arsenic sulphide is dissolved. A stream of chlorine is then passed into the solution until it is perfectly colourless: it is then gently heated for some time, and filtered from the small quantities of the oxides of copper and bismuth. Dissolve the residue in nitric acid, add an excess of ammonium carbonate, heat gently for some time, filter, redissolve in

T

nitric acid, and again precipitate the bismuth by the addition
of ammonium carbonate in excess, dry, and weigh the
bismuth as trioxide.   Mix the two filtrates and precipitate
the copper by means of caustic potash.

The filtrate containing the arseniate of potassium is acidified with hydrochloric acid, and ammonia, and 'magnesium
mixture' added.   The magnesium-ammonium arseniate may
be obtained highly crystalline (in which state it washes
better), by first adding the magnesia mixture to the acid
liquid, and then a large quantity of ammonia—about one-
fourth of the bulk of the entire liquid.   After standing 24
hours, pour the liquid on to a weighed filter, wash the precipitate with ammonia water until the washings acidified
with nitric acid give only a slight opalescence with silver
nitrate.   The magnesium-ammonium arseniate is dried in
the air-bath at 120°, and ignited at a low red heat in a
weighed porcelain crucible.   The ignited residue has the
composition $Mg_2As_2O_7$, corresponding to magnesium pyrophosphate.

To the rose-coloured filtrate containing the cobalt, nickel,
and iron, add nitric acid, and boil, and then solid sodium
carbonate, until the liquid is nearly neutralised, and becomes
slightly turbid, owing to the separation of ferric hydrate :
complete the precipitation by adding a solution of succinate
of soda or ammonia, wash and dry the precipitate, ignite it,
and weigh it as Fe_2O_3.

The nickel and cobalt are co-precipitated by adding
potash to the boiling solution ; the precipitate is filtered off
and washed, drained as far as practicable by the pump, the
filter spread on a plate, and the precipitate, detached as far
as possible by means of a thick platinum wire, transferred to
a small porcelain dish : the filter is re-folded and dropped
back again into the funnel.   Treat the precipitate in the dish
with dilute hydrocyanic acid, and then with potash solution,
and again with hydrocyanic acid, and warm on the water-
bath, until no further solution occurs.   A minute quantity
of substance frequently remains undissolved : it consists of

paracyanogen, and retains a small quantity of the mixed oxides. This is added to the small quantity remaining on the filter, which is now washed, dried, ignited, and weighed. The proportion of the two metals contained in it is calculated from the results of the after-separation. The solution of nickel and cobalt is boiled to expel the excess of acid : it is reddish-yellow, and consists now of cobalticyanide of potassium and double cyanide of nickel and potassium. Add to the hot solution precipitated (yellow) mercuric oxide, and

Fig 64.

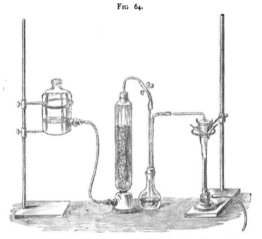

again boil for some time. All the nickel is precipitated, partly as cyanide, partly as sesquioxide, the mercury combining with the liberated cyanogen. Wash the precipitate thoroughly, dry it, and heat it for some time in a covered porcelain crucible to expel the excess of mercuric oxide. The filtrate is nearly neutralised with nitric acid, and a neutral solution of mercurous nitrate is added in excess, when cobalticyanide of mercury is precipitated; this is washed, dried, and ignited, in an open porcelain crucible, until the weight is constant : it has the composition Co_3O_4.

By way of control it may be reduced to the metallic state, by heating in a stream of dry hydrogen. When the decomposition is apparently finished, allow the metal to cool in a current of the gas and weigh: again heat in hydrogen and again weigh, repeating the operation until the weight is constant. The apparatus employed for the reduction is seen in fig. 64. The crucible lid is pierced with a hole and provided with a porcelain tube.

XXXIII. FAHL-ORE (TETRAHEDRITE)

Consists of a mixture of sulphantimonites and sulpharsenites of copper, silver, iron, zinc, and mercury. Every specimen of the ore, however, does not contain all these substances: in some the silver, zinc, and mercury are entirely absent; in others the whole of the arsenic is replaced by antimony. The composition of the mineral may be represented by the general formula $\left. \begin{matrix} (M_2 : N'')\,4 \\ Sb_2 \quad : \quad As_2 \end{matrix} \right\} S_7$, in which M denotes Cu and Ag, and N'' denotes Fe, Zn, and Hg.

Determination of the Sulphur.—See Copper Pyrites, p. 209.

Determination of the Metals.—The decomposition of the finely-powdered mineral is best accomplished by heating it in a stream of dry chlorine. The apparatus required for this purpose is seen in fig. 64A. The substance is placed in the bulb-tube *a*; the second bulb serves to collect the greater portion of the sublimate, and thus prevents the narrower portion from being stopped up by the chlorides. The mineral is best weighed out from a tube of a diameter sufficiently narrow to allow of its being introduced into the bulb-tube. The bulb-tube is connected with the U-tube *b*, which is filled to the extent of $\frac{1}{3}$ of the two upper bulbs with a mixture of hydrochloric and tartaric acids. The flask A contains the chlorine mixture (1 part of salt, 1 of manganese dioxide, 1 of water, and 2 of sulphuric acid: the sulphuric acid and water are previously mixed, and allowed to cool); the small flask *c* contains strong sulphuric acid, and the U-tube *d* sulphuric acid and pumice. Allow this portion of the apparatus to become filled with chlorine before joining it to the bulb-tube, and do not heat the mineral until the gas is apparently without further action upon it. Indeed as soon as the chlorine comes in contact with the fahl-ore immediate signs of decomposition ensue, and the bulb becomes very hot. When it cools, gently heat it in order to drive off the

FIG. 64A.

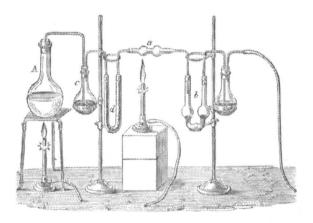

volatile products into the second bulb. Care must be taken to send only a gentle stream of chlorine through the apparatus, otherwise there is danger of a portion of the sublimate passing through the liquid unabsorbed. As soon as reddish vapours of ferric chloride make their appearance, discontinue the heating, and allow the apparatus to cool. Divide the bulb-tube *a* by means of a file between the two bulbs, and allow the portion containing the sublimate to remain in a damp place for 24 hours, in order that it may take up moisture from the atmosphere : this prevents the evolution of the great heat and consequent possibility of loss by vola- tilisation, when the chlorides are subsequently treated with water. In the meantime proceed with the analysis of the fixed residue contained in the first bulb; this contains the silver, copper, zinc, and a portion of the iron. It is placed in a beaker and digested with dilute hydrochloric acid until everything is dissolved, with the exception of the silver chlo- ride. This is filtered off, washed with hot water, dried, and weighed. The copper in the filtrate is precipitated by sul- phuretted hydrogen, re-dissolved in nitric acid, and weighed as oxide. The solution containing the zinc and iron is boiled with nitric acid for a short time, and the metals separated as in No. XIII. Part II.

The solution of hydrochloric and tartaric acids is poured into a beaker, the tube rinsed out, and the bulb containing the sublimate added. This solution contains the arsenic, anti- mony, mercury, and the remainder of the iron ; if it is cloudy, from the separation of a little antimony, warm it *gently*: some- times the turbidity is due to the separation of sulphur ; in that case the liquid should be filtered. The solution is heated to about 60°, and a current of sulphuretted hydrogen is passed through it for some time until the fluid smells strongly of the gas. It is allowed to stand for 12 or 15 hours, filtered, and washed with sulphuretted hydrogen water. The re- mainder of the iron is found in the filtrate ; it is precipitated by ammonium sulphide, washed with sulphuretted hydrogen water, re-dissolved in hydrochloric acid, the solution boiled with a little nitric acid or a few crystals of potassium chlo- rate, and the iron re-precipitated as ferric oxide by ammonia, washed, dried, and weighed.

The mixed sulphides of antimony, arsenic, and mercury are treated with potassium sulphide until the residue of mercuric sulphide is quite black. This is filtered off through a weighed filter, washed with water containing potassium sulphide, then with pure water, two or three times with alcohol, and finally with bisulphide of carbon, until a drop of the filtrate evaporated on a watch-glass leaves no residue. The filter is once more dried at 100° and weighed.

The sulphide of potassium solution is transferred to a porcelain crucible and saturated with chlorine gas. The solution is heated on a water-bath, and mixed with a considerable excess of hydrochloric acid and evaporated to about half its bulk ; an equal volume of hydrochloric acid is again added to the solution, again evaporated to one-half. A freshly prepared solution of sulphuretted hydrogen is next added (in the proportion of 100 c.c. of solution to each ·1 gram of antimonic acid supposed to be present), when antimony pentasulphide separates out. The excess of sulphuretted hydrogen is quickly expelled by blowing a stream of air through the liquid, and the precipitate is brought upon a weighed filter, drained by means of the pump, and washed six or eight times with alcohol, then with carbon bisulphide, and again with alcohol. The antimony pentasulphide is dried at 110°, and weighed.

The aqueous filtrate is mixed with a few drops of chlorine water, heated on the water-bath, and treated with a stream of sulphuretted hydrogen, and after standing for 24 hours in a warm place the precipitated arsenic pentasulphide is transferred to a weighed filter, and washed with alcohol and carbon bisulphide, as in the case of the antimony sulphide, and treated at 110° until its weight is constant.

Bournonite ($2PbSCuS.Sb_2S_3$); Boulangerite ($3PbS.Sb_2S_3$); red silver ore, $3Ag_2SAs_2(Sb_2)S_3$, and nickelspeiss may also be analysed by decomposition in a stream of chlorine.

XXXIV. DETERMINATION OF SILVER IN SOLUTIONS.

An excess of a standard solution of sodium chloride is mixed with a determinate volume of the silver solution to be tested, a few drops of potassium chromate solution are added, and the excess of the chlorine in solution is determined by adding a solution of silver of known strength until the orange colour of the silver chromate is persistent (compare No. I.

Part III.). This method is especially applicable to the determination of the strength of the silver solutions employed in
photography.

Pisani's Method (particularly applicable to the Estimation of Silver in Alloys and Ores).

If a solution of the blue compound of iodine and starch
be added to a neutral liquid containing silver nitrate, the
colour is rapidly destroyed, the iodine combining with the
silver to form silver iodide (and iodate ?). Immediately the
silver is completely precipitated the iodised starch solution
colours the liquid permanently blue, and thus marks the
completion of the process.

This method is, of course, only accurate in the absence of
substances, other than silver, which effect decomposition of
the blue solution. Tin, arsenic, and antimony, mercury, iron,
and manganese protoxides and gold must accordingly be
absent. Copper and lead do not influence the reaction.

Weigh out about 2 grams of iodine and 15 grams of pure
starch into a porcelain mortar, add 6 or 8 drops of water,
and mix intimately; transfer the mass to a flask, and heat it,
well closed, on the water-bath for an hour. The violet-blue
colour of the mixture will now have changed to dark greyish-
blue. Dissolve in water and dilute considerably.

To ascertain the value of the deep bluish-black solution,
transfer 10 c.c. of a neutral solution of silver nitrate, containing 1 gram of silver per litre, to a beaker, add a small quantity of pure precipitated calcium carbonate, and run in the
solution of iodised starch, with constant stirring, until the
yellow colour of the liquid (due to the silver iodide formed)
changes to greenish-blue. The 10 c.c. of silver solution
should require about 50 c.c. of the iodised starch solution.
The object of the calcium carbonate is to neutralise the
nitric acid liberated : it also renders the completion of the
reaction more distinctly visible.

The minute quantities of silver contained in lead ores and

in refined lead may be readily estimated by this method.
The nitric acid solution, prepared as directed on p. 259,
is mixed with sulphuric acid to remove the lead, the
liquid is filtered, mixed with calcium carbonate in excess,
and again filtered. A small additional quantity of the calcium carbonate is added to the filtrate, which is titrated with
the standardised iodine solution in the manner above described.

A determinate volume of the standard iodine solution,
p. 156, mixed with clear starch liquor, and diluted to centi-
normal strength, may be employed with equally good results.
(Field.)

XXXV. ASSAY OF SILVER IN BULLION, COIN, PLATE, &c.

The method about to be described is that generally prac-
tised in the European mints : it was originally devised by
Gay-Lussac, and has been rigidly investigated by Mulder.
Probably no quantitative process is susceptible of such a
high degree of accuracy as the estimation of silver by the
' humid method,' as the process of Gay-Lussac is generally
termed, in contradistinction to the older process of cupellation.

We have already indicated the leading features of this
method in describing the process for exactly estimating the
strength of a standard hydrochloric acid solution (p. 131).
This very simple process is, however, complicated by the
following circumstance : If we add 1 eq. of silver nitrate to
1 eq. of sodium chloride, both dissolved in water, we should
expect that all the silver would be precipitated, and that we
should obtain no subsequent turbidity by the further addition either of salt or of silver solution. But in reality we
find that the addition of either of the solutions produces a
turbidity. This remarkable fact probably depends upon
the solvent action of the sodium nitrate, and upon the ex-
istence of a certain degree of equilibrium between the silver
nitrate and common salt, which is destroyed, with the im-

mediate formation of silver chloride, by the addition of either of the bodies.

It thus happens that if we add a decimal solution of salt (*vide infra*), drop by drop, to a solution of silver until no further turbidity is produced, and then add decimal solution of silver to the liquid, we again notice the formation of a slight precipitate. If we continue to add the decimal silver solution until the turbidity ceases, and once more add decimal salt solution, we shall again observe the formation of another precipitate. If we determine the number of drops required to pass from one limit to the other, we observe that the same number of each is needed. Suppose that we had added in the first place the salt solution until the exact point was reached at which no further turbidity was produced, and that we required to add 20 drops of the silver solution before the precipitate again ceased to form, we should find that it would be necessary to add 20 drops of the salt solution before this point of the non-formation of a turbidity was again reached. If we add exactly half this number of drops, viz., 10, we reach what Mulder terms the *critical point*, that is, the point at which both salt and silver produce equal precipitates.

We have, therefore, three different methods of determining the completion of the reaction : *a*. We may add the salt solution until the turbidity just ceases ; *β*. We may stop at the neutral point ; or *γ*. We may go back with silver solution, and continue the addition until no further turbidity is produced. Whichever method we adopt in standardising the solution of salt we must afterwards invariably employ. Thus we must not at one time end with salt and at another end with silver. Mulder has shown that the error by such a procedure amounts to 1 milligram per gram of silver : by employing first *a* and then *β* the difference amounts to 0·5 milligram at 16° C. Mulder has also shown that the degree of error varies slightly with the temperature and dilution of the liquids.

On the whole it is most convenient to adopt the first plan—
i.e. to continue the addition of the salt until no further precipitate is formed. We require for this method :

1. *Solution of Sodium Chloride.*—Dissolve 5·4145 grams of
salt in distilled water, and dilute to 1 litre. The temperature
of the solution should be 16° when measured. 100 c.c.
of this solution = 1 gram of silver. Call it 'Salt Solution
No. 1.'

2. *Decimal Solution of Sodium Chloride.*—Transfer 50 c.c.
of the above solution to a ½-litre flask, and dilute to the containing-mark with water at 16°. Call it 'Salt Solution No. 2.'

3. *Decimal Silver Nitrate Solution.*—Weigh out exactly
0·5 gram of pure silver (see p. 124) into a ½-litre flask, dissolve in 3 c.c. of pure nitric acid, and dilute with water at
16° to the containing-mark. 1 c.c. contains 1 milligram of
silver.

4. *Test Bottles.*—The bottles specially made for the method
should be procured. They are of white glass, about 250 c.c.
in capacity, and are fitted with accurately-ground stoppers,
the lower portion of which is pointed. On touching the
side of the bottle with the point of the moistened stopper, the
adhering liquid is readily detached. The bottles are placed
in well-fitting cases of cardboard or vulcanite, and when in
use are wrapped in a black cloth in order to protect the silver
chloride from the light.

We commence the process by determining the exact value
of the salt solution. Weigh out with the greatest possible
accuracy from 1·001 to 1·003 gram of pure silver, place it in
a test bottle, add 5 c.c. of pure nitric acid (sp. gr. 1·2), and
heat the bottle (placed obliquely) on the water-bath until the
silver is dissolved. From time to time blow out the nitrous
fumes from the bottle, occasionally shaking the liquid to
promote their expulsion. When solution is effected, allow
the bottle to cool for a short time and place it in water at a
temperature of about 16°. Remove it in about half an hour,
wipe it, and place it in its case. Transfer 100 c.c. of salt

solution No. 1, measured with the greatest care, to the bottle. Moisten the glass stopper with distilled water, insert it firmly in the neck, cover the bottle with the black cloth, and shake the whole violently for a minute or two, or until the silver chloride settles completely, leaving the fluid perfectly clear. Take out the stopper, rub it on the bottle to remove the adherent silver chloride, replace it, and shake down the silver chloride on the sides of the glass by giving the liquid a rotatory motion.   As soon as the chloride is deposited, again open the bottle, incline it, and allow ½ c.c. of decimal salt solution (Sol. No. 2) to flow in against the lower part of the neck of the bottle.   The salt solution should be added from a Mohr's burette, graduated into 0·1 c.c., and fitted with a glass stopcock. After the addition of the ½ c.c. of salt solution, raise the bottle from its case, and note the degree of turbidity, insert the stopper, and shake until the liquid is again quite clear. Repeat the agitation after each addition of the salt solution, and, as the turbidity decreases, add the solution in very small quantities : towards the end only two drops should be added at a time.   At this point read off the burette, and continue the addition of the salt solution by two drops at a time, reading off the burette after each addition, and agitating the liquid until no further precipitate is produced.   When the last two drops fail to produce a turbidity the process is at an end. The previous reading—that is, the one before the addition of the last two drops—is taken as the correct one.

If by mischance the exact point of the non-formation of a turbidity has been overstepped, add 2 c.c. of decimal silver solution, shake, and continue the addition of the salt solution until the proper point is again reached.

To take the most complicated case : let us suppose that we have weighed off 1·0023 gram of pure silver, added 100 c.c. of salt solution No. 1, and 4·2 c.c. of the decimal salt solution. We have reason to believe that we have overstepped the proper point, and we therefore add 2 c.c. of decimal silver solution, and again 1·8 c.c. of decimal salt, when turbidity ceases.

Amount of silver taken = 1·0023 + ·0020 = 1·0043 gram.

This required 100 c.c. of No. 1 salt solution + 6 c.c. of No. 2 salt solution, or altogether 100·6 c.c. of No. 1 solution for complete precipitation. Calculate the quantity required for 1 gram of silver :

$$1·0043 \; : \; 1·000 \; :: \; 100·6 \; = \; 100·17.$$

We thus find that 100·17 c.c. of the salt solution No. 1 exactly precipitates 1 gram of pure silver. It is desirable from time to time to repeat the determination of the strength of the salt solution.

For the actual assay, weigh out about 1·085 gram of standard silver (12·3 of silver to 1 of copper) in the test bottle, dissolve in 5 or 6 c.c. of nitric acid, add 100 c.c. of the salt solution No. 1, and proceed with the addition of the decimal salt solution as directed. Let us assume that we had weighed out 1·085 gram, and that it was necessary to add 6 c.c. of decimal silver solution before turbidity ceased : this, without sensible error, we may assume to be equal to 0·6 of solution No. 1. We thus find that 100·6 c.c. of the strong solution were needed to precipitate all the silver in the 1·085 gram of the alloy. But 100·17 c.c. of the solution were equal to 1 gram of pure silver. Therefore

$$100·17 \; : \; 100·6 \; :: \; 1 \; = \; 1·0043 ;$$

and accordingly 1,000 parts of the alloy would contain

$$1·085 \; : \; 1000 \; :: \; 1·0043 \; = \; 925·6 \text{ parts of pure silver.}$$

In the case of alloys of which the composition is not approximately known it is necessary to determine it by a preliminary trial before the regular assay is made. Weigh off from 0·5 to 1 gram of the alloy according to its richness in silver, dissolve in 4 or 5 c.c. of nitric acid in the usual manner, and add sodium chloride solution No. 1 from a burette, until no further precipitate is formed. Then calculate the amount of the alloy which will contain 1·002 gram of silver, and proceed with the assay in the manner directed.

In the case of alloys containing sulphur and gold, digest with the least possible quantity of nitric acid, blow out the nitrous fumes from the bottle, add strong sulphuric acid, and boil until the gold separates completely, and proceed with the assay in the usual way.

XXXVI. ASSAY OF GOLD.

For a description of the most accurate methods of assaying alloys of gold by cupellation, we would refer the student to Professor Jevons' excellent article on the subject in Watts' 'Dictionary of Chemistry,' vol. ii. p. 932. Longmans: 1869.

XXXVII. SEPARATION OF GOLD, SILVER, AND COPPER.

I. When the amount of silver in the alloy does not exceed 15 per cent., the whole of the gold and copper may be dissolved out by means of nitro-hydrochloric acid, the silver remaining as silver chloride. The solution of the finely-divided alloy after complete decomposition is evaporated nearly to dryness to expel the greater portion of the free acid, water is added, the solution filtered, and the silver chloride is washed, dried, and weighed. To the solution oxalic acid is added, whereby the gold is completely reduced to the metallic state. After standing for about 48 hours, the liquid is filtered, and the gold washed, dried, and weighed. The copper in solution is precipitated as sulphide by means of sulphuretted hydrogen : it may either be weighed as such, after covering it with sulphur and heating in a stream of sulphuretted hydrogen, or be converted into oxide by re-solution and precipitation with sodium hydrate.

II. If, on the other hand, the amount of gold in the alloy does not exceed 15 per cent., the whole of the silver is dissolved by prolonged boiling with moderately-concentrated

nitric acid.* The liquid is evaporated nearly to dryness,
water added, the silver precipitated by hydrochloric acid,
and the copper in the filtrate by sulphuretted hydrogen.
The gold after weighing is dissolved in cold nitro-hydrochloric acid, to ascertain that it is perfectly free from silver.
If any silver chloride is found it must of course be filtered
off and weighed.

III. By heating alloys of gold, silver, and copper with
strong sulphuric acid, the metals may be separated, whatever
may be their proportion. The finely-divided alloy is heated
with the concentrated acid until no more sulphur dioxide is
evolved, and the acid begins to volatilise. Water is then
added and the liquid boiled, and the silver and copper
separated as above. It is advisable to treat the gold again
with the acid before finally weighing it : if any additional
silver and copper are obtained their weight must of course
be added to the main quantities of these metals.

XXXVIII. Estimation of Mercury.

In ores and compounds containing mercury the amount
of the metal may be readily determined by heating the substance with quicklime. The process is conducted in the
apparatus represented in fig. 65. Into a piece of combustion tube, about 50 centimetres long and rounded at the
end, introduce a layer, 5 centimetres in length, of powdered
magnesite (*a*). Weigh out about 5 grams of the substance
into a glazed porcelain mortar and mix it intimately with
quicklime. Introduce the mixture into the tube (*a* to *b*),
and rinse out the mortar with a fresh portion of quicklime
(*b* to *c*) ; fill up the rest of the tube to within a few centimetres of the end with powdered quicklime (*c* to *d*), and draw
out the tube before the blow-pipe in the manner represented
in the cut. Gently tap the tube so as to make a channel for

---

* The nitric acid must not be too strong, otherwise more than traces
of the gold would be dissolved by the nitrous acid formed.

the gas, and place it in the combustion furnace, and immerse the drawn-out end just beneath the surface of water contained in a small flask. Heat the tube from *d* to *c*, at first gently, and then to bright redness, and gradually heat the portion of

FIG. 65.

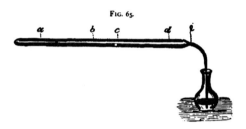

the tube containing the mercurial compound to redness. When the substance is completely decomposed, heat the magnesite to expel the traces of mercurial vapour within the tube. Whilst the tube is still red hot, cut off the tube at *e*, and wash any adhering mercury into the flask. Agitate the mercury beneath the surface of the water so as to bring it together into one globule, and in about half an hour decant the clear water ; pour the mercury on to a small weighed watch-glass, remove the water as far as practicable by filter-paper, and before re-weighing it place the watch-glass and metal for an hour or two beneath a bell-jar containing strong sulphuric acid. The only mercurial compound which thus resists complete decomposition is the iodide ; it may be readily analysed, however, by substituting copper filings for the lime. Cinnabar may be readily analysed by heating it with nitric acid of sp. gr. 1·4 in a closed tube for a couple of hours : the sulphur is converted into sulphuric acid, which may be determined as barium sulphate. Silica, heavy spar, &c., remain undissolved. The mercury is precipitated as calomel by phosphorous acid—as described below.

In the case of very poor ores of cinnabar containing large

quantities of bituminous matter, such as those of Austria, the foregoing process may be judiciously modified by first extracting the organic matter from the powdered ore by means of benzol, thoroughly drying the residue, and treating it with a solution of barium sulphide (containing about 50 grams of barium to the litre). The mercuric sulphide is precipitated from the solution by means of hydrochloric acid, filtered, dried, and digested with carbon bisulphide to remove admixed sulphur. The impure mercuric sulphide is then dried, re-dissolved in aqua regia, the liquid concentrated, and mixed with solution of phosphorous acid, allowed to stand about 12 hours in the cold, and the precipitated calomel collected on a weighed filter, washed with hot water, and dried at 100°.

Mercury may be readily determined by electrolysis. The solution of the metal is poured into a weighed platinum dish and slightly acidified with sulphuric acid, and the dish is connected with the zinc pole of a battery of six bichromate cells, the carbon end being attached to a piece of platinum foil which dips into the solution. Mercurous chloride gradually separates, and this is finally reduced to the metal, and in about an hour all the mercury is precipitated. The metal is washed with water, alcohol, and ether, and dried over oil of vitriol and weighed.

[*Note on the Preparation of Phosphorous Acid.*—Small pieces of phosphorus are introduced into a saturated solution of copper sulphate contained in a corked flask; metallic copper is first reduced, which is ultimately converted into black copper phosphide, and the acid solution consists of a mixture of phosphorous and sulphuric acids. The latter may be removed by the cautious addition of baryta water. The precipitate is allowed to settle, and the clear solution of phosphorous acid decanted and preserved in a well-stoppered bottle, as it oxidises on exposure to air. It is a very powerful reducing agent, and may be usefully applied in a variety of cases.]

XXXIX. COAL.

In the proximate analysis of coal we require to determine the amount of moisture, volatile matter, coke, ash, and sulphur. If the actual amount of carbon, hydrogen, and nitrogen is needed, recourse must be had to elementary organic analysis.

Determination of the Moisture.—About 2 grams of the finely-powdered coal are weighed out and dried between watch-glasses at 105–110° for an hour, and the loss is set down as moisture. With the quantity of coal taken the loss of weight appears to be greatest at the end of this time: on further heating, it actually increases in weight, an effect due probably to the oxidation of the finely-divided pyrites.

Determination of the Volatile Matter.—About 2 grams of the powdered undried coal are heated for four minutes over a Bunsen flame, and then immediately, without cooling, for the same length of time over the gas blow-pipe flame. The loss is set down as volatile matter + moisture. The residue gives the coke + ash.

Determination of the Ash.—From 3 to 5 grams of the finely-divided coal are heated in a small platinum dish over a Bunsen lamp. Usually the incineration proceeds with rapidity: if it is found necessary to increase the draught of air over the heated mass, the arrangement described in the section on Analysis of Ashes of Plants may be employed.

Determination of the Sulphur.—The sulphur in coal exists in two modifications—as pyrites and as calcium sulphate. The sulphur contained in the pyrites alone influences the economical application of the fuel.

The total amount of sulphur may be determined by heating about 2 grams of the powdered coal with four times its weight of pure sodium carbonate in a platinum dish, or the coal may be heated in a current of oxygen and the gases passed through a solution of hydrochloric acid and bromine, the sulphuric acid being determined by precipitation with barium chloride. The sulphur existing as calcium sulphate may be determined by boiling 5 grams of the finely-powdered coal with a solution containing about the same weight of pure sodium carbonate, whereby the calcium sulphate is decomposed, calcium carbonate and sodium sulphate being formed. Filter the solution, acidify with hydrochloric acid, and add barium chloride. The difference between the total amount of sulphur and that found after boiling with sodium carbonate represents the amount as pyrites. The same process is of course applicable to the determination of the iron sulphide and gypsum in coke.

Determination of Specific Gravity.—It is occasionally desirable to ascertain the weight of a cubic foot of the fuel, or the number of cubic feet corresponding to a ton. This is easily calculated when the specific gravity of the coal is known. The specific gravity is readily determined by weighing the coal in air; and in water by suspending it from the arm of the balance by a hair or thin wire. The piece taken should not be too small, and care should be taken that no air bubbles adhere to it during the weighing. It is desirable, too, that the coal be soaked sufficiently : this is easily effected by immersing the lump, after attaching the hair or wire to it, in water, in the flask of the filter-pump, and exhausting the air within the apparatus as far as practicable. The weight of a cubic foot of the coal in pounds is found from the expression :

$$\text{log. sp. gr.} + 1\cdot79588 = \text{log. wt. of cb. foot.}$$

The number of cubic feet in a ton

$$= 1\cdot55437 - \text{log. sp. gr.} = \text{log. cb. ft.}$$

XI. EXAMINATION OF WATER USED FOR ECONOMIC
AND TECHNICAL PURPOSES.

1. *Collection of the Sample.*—The water to be analysed should be collected in stoppered glass bottles—those known as 'Winchester Quarts,' which hold about $2\frac{1}{2}$ litres, may be conveniently employed. As a rule two of these bottles will contain sufficient water for an ordinary examination. If, however, an exhaustive analysis is required, double or even treble this amount may be necessary. Care must be taken

that the bottles, and the vessels employed to fill them, are quite clean, and a due amount of judgment should be used to obtain a representative sample.   In collecting the water from a river or tank the bottles themselves should be immersed below the surface, and rinsed once or twice with the water.   In taking the water from a pump or pipe a considerable quantity should be allowed to flow away before the sample is collected.   The bottles should be filled up nearly to the neck and the stopper tied down with string : no luting or sealing-wax should be used.

Glass bottles are preferable to stoneware jars, for the reason that earthenware is not readily cleaned ; moreover it is liable to affect the hardness of the water, as the clay not unfrequently contains notable quantities of calcium sulphate.

2. *Preliminary Observations.*—Fill a tall narrow cylinder of white glass with the water to be examined, and compare its *colour* with that of distilled water contained in a similar cylinder.   Heat a portion to about 30° in a wide test-tube, shake, and note if the water possesses any peculiar *odour* or *taste.*

In the outset the analyst must decide whether the water for analysis is to be filtered or not.   His decision will depend upon the manner in which the water is used by the consumer.   If it be considered necessary to filter the sample, care must be taken that the paper employed is free from ammonia.   It should be steeped in distilled water for some time before use, dried and folded, and heated in a weighed tube for some hours at 120°.   It is placed in the desiccator and weighed when perfectly cold.   It is now properly fitted into the funnel, and the filter-flask is replaced by a clean ' Winchester Quart,' in which the filtrate is directly received. The quantity of the water to be filtered is measured ; as soon as the whole has passed through, the funnel is removed from the bottle and washed with distilled water (the washings must not be allowed to mix with the filtered water), again dried at 120° for some hours in the stoppered tube, and again weighed. The increase in the weight gives the quantity of *total suspended matter* in the known volume of the water. Incinerate the paper in a small platinum crucible, treat the residue with a few drops of solution of ammonium carbonate, dry and weigh ; the quantity in excess of that contained in the filter gives the amount of *suspended inorganic matter* in the water.

If it is considered unnecessary to filter the water, care should be taken to shake the bottle before withdrawing portions for analysis.

3. *Estimation of the Ammonia.*—It is desirable to proceed at once with the determination of this constituent, since it is the most liable to change. The method of estimation is based upon the fact that an alkaline solution of mercuric iodide added to a liquid containing ammonia produces a brown colouration, due to the formation of the iodide of tetramercurammonium. This test, known as Nessler's, is capable of detecting 1 part of ammonia in 20,000,000 parts of water.

Preparation of the Nessler Test.—Dissolve 35 grms. of potassium iodide in 120 c.c. of water, transfer 5 c.c. of the solution to a clean beaker, and add, little by little, a cold concentrated solution of mercuric chloride to the remainder until the mercuric iodide ceases to be re-dissolved on stirring. Add the 5 c.c. of the potassium iodide to re-dissolve the remaining mercuric iodide, and cautiously continue the addition of the corrosive sublimate solution until a very slight precipitate only remains. Now add an aqueous solution of potash, prepared by dissolving 100 grams of ' stick ' potash in 200 c.c. of water, and dilute the mixture to 500 c.c. The liquid should be allowed to stand for a short time, and a portion decanted into a small bottle for use. As the small bottle becomes empty it is replenished from the other. In addition we require :

(α) *A Standard Solution of Ammonium Chloride.*—Dissolve 0·7867 grm. of pure ammonium chloride in a litre of distilled water. Pour it into a clean stoppered bottle, and label it 'Ammonium Chloride Solution No. 1.' Withdraw 100 c.c. of this solution and dilute it also to 1 litre. Call it 'Ammonium Chloride Solution No. 2.' 1 c.c. of the latter solution contains ·025 *of a milligram of ammonia.* The solution should be delivered from a Mohr's burette, fitted with glass stopcock, and graduated to tenths of a cubic centimetre.

(β) *A small Pipette to deliver about* 1 *c.c. of the Nessler Test.*—This may readily be made from a short piece of glass tube. Also several cylinders, marked A, B, C, D, &c., about 20 cm. in height and of 60 c.c. capacity. To graduate them, transfer 50 c.c. of water to each, and mark the level of the liquid on the glass. Also two or three pieces of thin glass tube, about 30 cm. in length, and 3 mm. in external diameter, the ends of which should be blown into bulbs of such diameter that they will readily pass into the cylinders; the other ends are sealed.

(γ) *Distilled Water free from Ammonia.*—The distilled water of the laboratory must be tested for ammonia. Rinse one of the cylinders with the water and fill it up nearly to the top; add 1 c.c. of the clear Nessler solution, and agitate with the bulb-tube (β) by drawing it up and down a few times within the cylinder. If, after standing for five minutes, the water remains perfectly uncoloured, it may be considered free from ammonia. If it shows a yellow or brown tint, re-distil it after addition of about 1 gram of pure sodium carbonate; collect the distillate in a Winchester quart, as soon as 50 c.c. received in one of the cylinders gives no reaction for ammonia on testing with the Nessler solution. If ordinary water is used, the distillation must not be carried to dryness, and the water remaining in the retort or boiler must be thrown away before a fresh quantity is distilled.

The Process.—Transfer 50 c.c. of the natural water to be tested to one of the glass cylinders standing on a sheet of white paper, add 1 c.c. of the Nessler solution and agitate with the bulb-tube. Run 50 c.c. of distilled water into a second cylinder, add ·2 c.c. of Ammonium Chloride Solution No. 2, mix thoroughly, and compare the tints in the two cylinders. If they are about equal in intensity, take half a litre for the estimation : if the coloration in the natural water is the more intense, take a proportionately less quantity. This testing is simply preliminary : its object is to afford an idea of the proper quantity to take for the actual estimation. Observe whether the natural water becomes turbid after the addition of the Nessler test : a decided precipitate is due to lime or magnesia salts, and indicates hardness. 500 c.c. of the water, or a less quantity if the preliminary testing has shown that ammonia is present in considerable amount, are placed in a capacious retort, and connected with a Liebig's condenser, which should be freed from ammonia by previously blowing steam through it for a few minutes. If less than 500 c.c. of water have been taken, the liquid in the retort should be made up to this volume by the addition of pure distilled water before the distillation is commenced. Add about 1 gram of recently-heated and pure sodium carbonate, note if much precipitate is formed, and distil rapidly over the direct flame. Collect 50 c.c. of the distillate in one of the cylinders, A ; when filled replace it by a second cylinder, B. When the second cylinder is full, remove the lamp ; add 1 c.c. of Nessler's solution to the *second* 50 c.c. of the distillate (*i.e.* B), agitate, and place it on a sheet of white paper. Now fill up a third cylinder, Z, with pure distilled water to within a few centimetres of the level of that in B, add as much standard Ammonium Chloride Solution No. 2 as you think will produce the same depth of colour as in B, and afterwards 1 c.c. of Nessler's solution : add a little distilled water if necessary, so as to make the two levels in the tubes coincident. Agitate and compare the tints. If the colour of the liquid in the two cylinders, after standing about 5 minutes, is equal, we at once know the amount of ammonia contained in B : it is equal to that contained in the volume of standard ammonium chloride solution added to Z.   If the intensity in Z is not equal to that in B, pour away the contents of the former cylinder, rinse it, fill it with a second portion of distilled water, add more or less ammonium chloride solution, as the case may be, and 1 c.c. of the Nessler test.[*]

If the quantity of ammonia in B does not exceed ·o1 of a milligram (equal to o·4 c.c. of the standard ammonium chloride solution), the distillation may be discontinued : if the amount is greater than this, the boiling must be renewed, and successive portions of 50 c.c. of the distillate tested until the above limit is reached.   If the quantity of ammonia in B does not exceed that corresponding to o·8 c.c. of the standard solution of ammonium chloride, the amount in A may be determined in the manner directed ; if the quantity is greater than this, 25 c.c., or less if need be, of the solution must be transferred to another cylinder, diluted to 50 c.c. with pure distilled water and tested as above.   A colour produced by more than 4 c.c. of the ammonium chloride solution cannot be conveniently compared, since the liquid is apt to become turbid.   In the case of waters known to contain much ammonia, as in sewage, distil over 100 c.c. at once into a larger cylinder, withdraw an aliquot portion, dilute to 50 c.c., and titrate in the manner directed.

The determination of the ammonia in water used for drinking is of great importance, since an undue proportion of this substance denotes contamination with sewage. Sewage may contain from 2 to 10 parts of ammonia in 100,000 parts of liquid : river-waters may be said to contain on the average about o·o1 part, although this amount is

---

[*] The addition of more ammonium chloride solution *after* the Nessler test has been mixed with the liquid would cause a turbidity, which prevents accurate comparison.

subject to great variation. Bad well-waters sometimes contain as much as 0·5 to 1 part in 100,000 parts.

Estimation of ' Albuminoid Ammonia.'—Messrs. Wanklyn and Chapman have found that many nitrogenous organic substances yield either the whole or a definite portion of their nitrogen in the form of ammonia when boiled with an alkaline solution of potassium permanganate. Hippuric acid parts with all its nitrogen as ammonia when thus treated : whereas albumen gives up only 10 per cent. of ammonia, uric acid 7 per cent., and creatine 12·6 per cent. Since it is highly probable that the azotised organic matter contained in water is of an albuminoid nature, its quantity may be approximately estimated by determining the quantity of ammonia yielded by boiling the water with an alkaline solution of potassium permanganate ; according to Messrs. Wanklyn and Chapman, 'the disintegrating animal refuse' in the water 'would be pretty fairly measured by ten times the albuminoid ammonia which it yields.' *

The liquid remaining in the retort after distillation with sodium carbonate (Estimation of Ammonia) is mixed with 50 c.c. of a solution obtained by dissolving 8 grams of potassium permanganate and 200 grams of potassium hydrate in 1 litre of distilled water. The mixture should be boiled for some time previous to use and preserved in a well-stoppered bottle. After the addition of the permanganate solution, heat the retort over the naked flame, and distil successive portions of 50 c.c., and determine the quantity of ammonia present in them with the Nessler solution. As soon as the distillate contains less than ·01 milligram of ammonia, the process may be considered at an end. Add together the several quantities of ammonia obtained. The succussive ebullitions occasionally noticed in bad water may be diminished by throwing a number of recently-ignited pieces of pumice into the liquid.

* Wanklyn and Chapman, 'Water Analysis,' p. 68.

Estimation of Organic Carbon and Nitrogen (Frankland and Armstrong's Process).

Drs. Frankland and Armstrong have proposed to estimate the carbon and nitrogen contained in the organic matter present in water by direct combustion. The water is evaporated to dryness, and the residue is mixed with cupric oxide and burnt as in the elementary analysis of an organic compound. The resultant gas is collected over mercury, and the proportion of carbon dioxide and nitrogen determined by gasometric analysis. The process occupies considerable time, but if the evaporation of the water be commenced as soon as the ammonia-determination has been made, the residue will be ready for the combustion (which occupies about an hour) by the time that the hardness, total soluble matter, nitrates, &c., have been estimated.

1. *Evaporation of the Water.*—The quantity of the water needed for analysis will depend upon its quality. If less than 0·05 part of ammonia in 100,000 parts of water has been found, a litre should be taken; if more than 0·05, but less than 0·2, half a litre will suffice; if more than 0·2, and less than 1·0, a quarter of a litre should be used. Of sewage, which is much richer in organic carbon and nitrogen, 100 c.c., or even less, may be taken.

Transfer the measured quantity of the water to a large flask, add to it 20 c.c. of a saturated solution of washed sulphurous acid, and, if it does not exceed 250 c.c., boil the mixture for a few seconds to expel the carbon dioxide present. Transfer the water, little by little, to a hemispherical glass dish, 10 centimetres in diameter, and shaped somewhat like a finger-bowl: during the evaporation the glass dish should be supported in a copper basin provided with a projecting flange and resting on the water-bath, and over it should be placed a glass shade, about 12 in. high (such as is used for covering statuettes). The steam condenses in the inside of the shade and flows down into the copper dish, filling the space between the two dishes. The excess of water flows out by a small lip on the edge of the copper dish, and is led off by a piece of tape. The destruction of the nitrates and nitrites by the sulphur dioxide may be greatly accelerated by the addition of 2 or 3 drops of ferrous chloride solution (prepared by dissolving well-washed ferrous hydrate precipitated from ferrous sulphate by pure soda solution in the minimum quantity of pure hydrochloric acid) to the first dishful of the water; if the water is free from carbonates it will be necessary also to add 1 or 2 c.c. of a solution of sodium bisulphite in order to combine with the sulphuric acid formed, which if free would decompose the organic matter on concentration. If, however, the water in the glass dish or flask ceases at any time during the progress of the evaporation to smell of sulphur dioxide, more of the solution should be added. If the water is found to contain much nitric acid it may be necessary to digest the residue with a dilute solution of sulphur dioxide, and again evaporate to dryness, to ensure the complete elimination of the inorganic nitrogen.

2. *Combustion of the Residue.*—Introduce a small quantity of cupric oxide in fine powder (made by oxidising the metal in air*) into the dish, and mix it thoroughly with the residue by the aid of a small steel spatula : this should be very elastic, so that by accommodating itself to the curvature of the dish the dried residue may be completely detached from the glass. Fill about 3 centimetres of the carefully-cleaned combustion tube (which should be about 40 centimetres in length and 1 centimetre in internal diameter, sealed at one end and rounded like a test-tube) with pure copper oxide, and transfer the whole of the mixture in the glass dish to the tube, rinsing the dish with small successive portions of

---

* The copper oxide may be prepared by cutting copper wire or sheet into small pieces, washing the metal, and heating it in a muffle. The oxide obtained by strongly heating the copper nitrate cannot be well employed, as it is very difficult to expel the last traces of nitrogen from it. The cupric oxide remaining in the tube after the combustion (with the exception of that with which the substance was mixed) may be used again after it has been re-heated in a current of air.

FIG. 66.

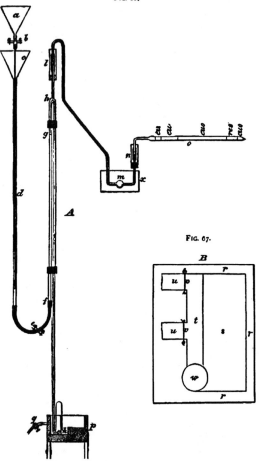

FIG. 67.

pure cupric oxide in fine powder. Add copper oxide to the
tube until it is a little more than half-filled, insert a cylinder
of metallic copper, about 8 centimetres in length, made by
wrapping fine copper gauze round a piece of thick copper
wire,* and then add 2 centimetres of granular copper
oxide in order to oxidise any carbon monoxide which
might be formed on burning. The end of the tube is
softened in the flame of the blow-pipe and drawn out to
form a tube about 150 millimetres long and 4 millimetres in
diameter. Bend the tube at right angles, fuse the edges in
the flame, place it in the combustion furnace and attach it
to the Sprengel pump.

Fig. 66 shows the arrangement of this apparatus as applied
to the purpose of exhausting the combustion tube. *a* is a
glass funnel maintained full of mercury, and connected by
means of a short piece of caoutchouc tube, on which is a
screw clamp, *b*, with a long narrow tube which passes nearly to
the bottom of a wider tube, *d*, 90 centimetres in length and
about 1 centimetre in internal diameter ; the upper end of
d is connected with a glass funnel in the manner represented
in the figure. *d* is connected with the tube *f g* by a piece
of strong caoutchouc tube covered with tape and provided
with a screw clamp. The tube *f g* is about 6 millimetres in
diameter and 600 millimetres in length; it is attached to a
tube, *g h k*, about 1,500 millimetres long and 6 millimetres in
external diameter, but with a bore of only 1 millimetre. The
portion *g h* is about 20 centimetres long ; the portion *h k* is
about 130 centimetres. To give them stability the tubes are
fastened together by caoutchouc and copper wire. At the
upper portion of the bend is a tube, *h l*, about 12 centimetres
long and 5 millimetres in diameter. The combustion tube

---

* The cylinder must be previously heated in the lamp, so as to
oxidise it superficially : it is then placed in a tube and heated in a
current of hydrogen, in order to reduce the oxide formed. The
cylinder must be allowed to cool in the gas before it is withdrawn from
the tube.

o is connected with the tube *h l* by the tube *l m n*, of the same diameter as the tube *h k*. The tube *l m n* is connected with the tube *h l* and with the combustion tube by short lengths of well-fitting caoutchouc tube ; the joint at *l* is bound round with copper wire, and is surrounded with glycerine, contained in the wider tube supported by a cork on *h l*; the joint at *n* is in like manner surrounded by a wider tube filled with water. On the tube *l m n* is a small bulb, which is immersed in cold water during the combustion ; its object is to receive the greater portion of the water formed on burning the residue. The tube *h k* is re-curved at *k*, where it ends in the mercury-trough *p*. The trough *p* (shown in plan at B, fig. 67) is cut out of a solid piece of mahogany. It is 20 centimetres long, 15·5 centimetres wide, and 10 centimetres deep, outside measurement. The edge *r r* is 13 millimetres wide, and the shelf *s* is 65 millimetres wide, 174 millimetres long, and 50 millimetres deep from the top of the trough. The channel *t* is 25 millimetres wide, and 75 millimetres deep ; at one end of it is a circular well, *w*, 42 millimetres in diameter, and 90 millimetres deep. The recesses *u u* receive the re-curved ends of the Sprengel pumps : the object of having two recesses is to allow of two experiments being made simultaneously : each recess is 40 millimetres long, 25 millimetres wide, and 75 millimetres deep. The tubes destined to receive the gases are supported against the iron wires *v v*. The trough stands upon four short legs, and has a side tube and clamp, *q*, to draw off the mercury to the level of the shelf *s* when necessary.

When everything is arranged, heat the fore part of the tube containing the metallic copper and unmixed copper oxide, and allow a gentle stream of mercury to flow from the funnel *a* : on reaching *h* the metal passes down the tube *h k* in detached portions, each carrying before it a small quantity of air from the combustion tube. Care must be taken so to control the flow of mercury that it does not rise into the tube *l m n.* The bulb on the tube *l m n* is sur-

FIG. 68.

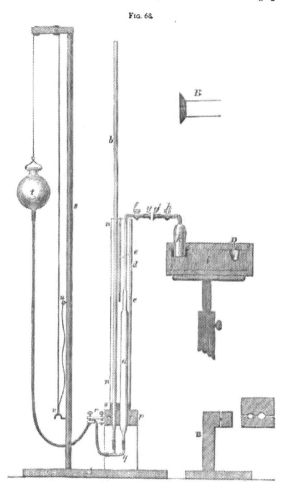

rounded by hot water during the exhaustion in order to expel any moisture which may remain in it from a previous experiment. If the fall is properly regulated the exhaustion will be complete in about ten minutes, when the mercury will be heard to fall with a sharp clicking sound. The action of the pump is now arrested ; a small tube filled with mercury is inverted over the end, k, of the tube, the hot water in x is replaced by cold water, and the rest of the tube is gradually heated to redness. In about an hour the combustion will be terminated. The pump is again set in operation and the gases are transferred to the tube.

Measurement and Analysis of the Gases.—The gases produced in the combustion consist of carbon dioxide, nitrogen, nitrogen dioxide, and occasionally, if the operation has been conducted too rapidly, sulphur dioxide and carbon monoxide. Their measurement and analysis may be conveniently made in the apparatus seen in fig. 68, which is essentially that devised by Frankland for the separation of gases incidental to water-analysis. $a\,c\,d$ is the measuring tube : the portion a is about 370 millimetres long and 18 millimetres wide, c is 40 millimetres long and 7 millimetres wide, and d is 175 millimetres long and 2·5 millimetres in diameter. To the upper end of d is attached a tube with capillary bore, bent at right angles and provided with a stopcock, f. The measuring tube is graduated from below upwards at intervals of 10 millimetres, the zero being about 100 millimetres from the lower end. The upper portion of d is divided into millimetres. Attached to the tube and stopcock f, is a steel cap, shown on a larger scale at B, fig. 68. The lower portion of a is drawn out until it is only about 5 millimetres wide : the tube b, which is about 1·2 metre long and 6 millimetres in internal diameter, is also narrowed at the lower end. Both a and b pass through the

---

* In newer forms of the apparatus Frankland has dispensed with the steel caps: the tube from the laboratory vessel being fitted by a cap of unvulcanised caoutchouc into a cup-shaped vessel attached to the capillary tube of the measuring apparatus.

caoutchouc stopper *o*, which is fitted into the glass cylinder *n n*, which is filled with cold water with the view of giving a definite temperature to the enclosed gas : this temperature is ascertained by a thermometer, *c*, suspended by a hook from the edge of the cylinder. Uniformity in the temperature of the mass of the water may be ensured by agitating it with an iron wire, the end of which is bent in the form of a ring. The tube *b* is graduated into millimetres, the zero being about 10 millimetres above the stopper *o*, and on a level with that of *a c d*. The tubes *a* and *b* are supported by the wooden clamp *p* (seen in end elevation and plan at B and c); the clamp is drawn together by two screws, the tubes being covered with caoutchouc where they fit into the holes to protect them from breakage. The clamp is supported by an upright piece of wood (seen in B) which is screwed into the base : *a* and *b* are connected by tubes of caoutchouc, covered with tape and bound with wire, to the tube *q*, which is also connected with the long caoutchouc tube leading to the glass reservoir *t*. This tube, which should have an internal diameter of not more than 2 millimetres, passes through the steel clamp *r*, the lower portion of which is fixed into *p*. The reservoir *t* is suspended by a cord passing over pulleys, in the arm of the iron rod *s*. On releasing the loop on the cord from the hook *v*, the reservoir sinks from about 10 centimetres above the level of the stopcock *f* to the level of the bottom of the clamp *p*. In the jar *k*, termed the laboratory vessel, the gases are

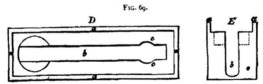

FIG. 69.

subjected to the action of absorbents. It is 100 millimetres high and 40 millimetres in internal diameter, and is fur-

x

nished with a capillary tube, glass stopcock, and steel cap, *g′ h*, exactly like *f g*. The mercury trough *l*, seen in plan in D and in section in E in fig. 69, is cut out of a solid piece of mahogany: it is 265 millimetres long, 80 millimetres broad, and 90 millimetres deep, outside measurement. The rim *a a* is 8 millimetres broad and 15 millimetres deep. The channel *b* is 230 millimetres long, 26 millimetres broad, and 65 millimetres deep. In the larger excavation at the end of the channel is placed the laboratory vessel : it is 45 millimetres in diameter and 20 millimetres in depth below the top of *b*. The small cavities *c c* are to receive the capsule employed to transfer the tube containing the gases from the trough of the Sprengel pump. The trough *l* rests on a telescope-table, and its height is so adjusted that when the laboratory vessel is placed in the cavity the faces of the steel caps are in exact coincidence. (Fig. 68.)

Before using the instrument the 'corrections for capillarity' must be determined. When the mercury in the tube *b* and in the measuring tube *a c d* is freely exposed to atmospheric pressure, it will be noticed that the levels of the metal in the two tubes are not coincident ; the level in *b* is slightly higher than in *a* ; on the other hand the level in *c* and *d* will be found to be higher than that in *b*. The difference in each portion should be determined by taking several observations : the correction will also include the error arising from difference of level in the zeros of the graduations of *b* and *a c d*. The determination of the levels should be made by the aid of a telescope sliding on a vertical rod.

To determine the capacity of the measuring tube at each graduation, first fill the entire tube with mercury, so that the metal drips from the cap *g*. Close the stopcock *f*, and slip a piece of caoutchouc tube over the cap ; attach the other end of the tube to a funnel filled with distilled water ; lower the reservoir *t*, and open the clamp *r* and the stopcock *f*. As the mercury flows into the reservoir, water is drawn through the capillary tube. As soon as it is below the zero on *a*, close *f*, remove the caoutchouc tube from the cap and slightly grease it, to allow water to pass through it without adhering. Raise the reservoir, open *f*, and expel the water until the upper portion of the mercury meniscus is coincident with the zero of the graduation. Now allow the water to flow out into a small tared flask until the level of the mercury is coincident with the next graduation, controlling the influx of the mercury by the clamp *r*. Read off the temperature of the water in the cylinder *n*, weigh the water in the flask, and calculate its volume from Table II. in the Appendix. Repeat the determination between successive graduations on the whole length of the tube in exactly the same manner.

A table, showing the capacity of the tube at various points, is then constructed, the intermediate graduations being obtained by interpolation : as the calculations are much facilitated by the use of logarithms, it will be found more convenient to set down in this table the logarithms of the capacities in place of the natural numbers.

To use the apparatus, grease the stopcocks *f* and *h* and the faces of the caps *g* and *g'* with a little resin cerate mixed with oil. Fill *a c d* with mercury and close *f*. Place the laboratory vessel in its cavity, and suck out the air as far as practicable by the aid of a caoutchouc tube, which is removed as soon as the jar is filled. Any remaining air may be drawn away by aspirating at *g'*. Close *h*, and fasten the faces of the caps tightly together by the aid of the clamp A. Of course the entire apparatus must be quite free from air,

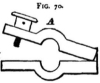

FIG. 70.

and on opening the stopcocks the mercury should flow freely through the capillary tubes. To determine if the several joints of the apparatus are air-tight, close *h*, and lower the reservoir, until it is on a level with *p*. Since the stopcocks and joinings are thus subjected to a pressure of nearly half an atmosphere, any imperfection which may cause leakage will be readily detected. After the trial replace *t* in its original position.

The gas to be analysed is decanted into the laboratory vessel and treated with one or two drops of strong potassium bichromate solution, to ascertain if it is free from sulphur dioxide. If this gas is absent the colour of the solution will be unaltered; if present a portion of the chromic trioxide will be reduced, and the liquid will become green. If any change is observed, pass up a few more drops of the solution, to complete the absorption of the gas. Open the stopcocks and lower the reservoir, and transfer the gas to the measuring tube ; close *h* so soon as the liquid in the laboratory vessel is within 10 mm. from the stopcock. The quantity of gas remaining in the capillary tube is too minute to affect the experiment. The apex of the mercury meniscus in *a c d* (as seen through the telescope) is made to coincide with the nearest division on the tube by allowing mercury to flow in from the reservoir *t*. Read off the levels of the mercury in tubes *b* and *a c d*; note the temperature of the water in *n*, together with the height of the barometer.

Pass a few drops of strong potash solution by means of a pipette into the laboratory vessel, and return the gas to it. The absorption of the carbon dioxide will be complete in about five minutes. The gas now consists of nitrogen mixed with a small quantity of nitric oxide ; it is again brought into the measuring tube, and its volume is ascertained in the same manner as before. If the volume of the gas is very small it is possible that it may already contain oxygen ; if so, any nitric oxide which might have been formed will have been converted into nitrogen tetroxide, which will have been absorbed by the potash solution. The volume of gas abstracted in this case is, however, too small to affect the result. To ascertain if oxygen is present, pass up a small quantity of a cold saturated aqueous solution of pyrogallic acid into the jar, and by *gently* shaking the stand of the trough, throw the liquid up against the sides of the jar in order to promote the absorption. As soon as the liquid runs down from the

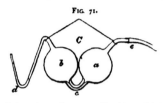

Fig. 71.

glass without the formation of a dark red stain, the absorption of the oxygen is complete. If oxygen be absent, it will be necessary to introduce a few bubbles of that gas in order to oxidise the nitrogen dioxide which may be mixed with the nitrogen. The addition of the oxygen may be conveniently made from the pipette shown in fig. 71. The bulbs *a* and *b* are about 5 cm. in diameter; the neck joining them is narrowed, so that mercury flows through it but slowly. To use the instrument fill the bulb *b* and the tubes *d* and *c* with mercury; introduce the tube *d* into a small tube containing oxygen, standing over the mercury trough, and gently aspirate by the aid of the caoutchouc tube *e*. A few bubbles are readily drawn over into *b*, and the gas is confined by the mercury in *d* and *c*; on introducing the limb *d* beneath the edge of the laboratory jar and gently blowing through *e*, the oxygen may be transferred to the gas under examination. Allow the mixed gases to remain in contact with the potash solution for a few minutes. When the nitrogen tetroxide and excess of oxygen have been absorbed, transfer the gas back again to the measuring tube and determine the volume of the residual nitrogen.

The three reduced volume-readings—1st, of the total gas (A); 2nd, of the nitrogen and nitrogen dioxide (B); and 3rd, of the nitrogen (C)—furnish all the data for obtaining the total volume of nitrogen and carbon dioxide in the gaseous mixture.

$$A - B = \text{vol. of } CO_2.$$
$$\frac{B + C}{2} = \text{vol. of N.}$$

From the corrected volumes the weights of the carbon and nitrogen are readily calculated.

The calculation, as Dr. Frankland has pointed out, may be simplified by considering the original gaseous mixture as nitrogen, so far as volume-weight is concerned. If A be the weight of the total gas, B its weight after treatment with potash, and C after absorption by pyrogallate, the weight of carbon will be $\frac{3}{7}$ (A—B), and the weight of nitrogen $\frac{B+C}{2}$ since the weights of carbon and nitrogen in equal volumes of carbon dioxide and nitrogen, measured under the same conditions, are as 6 : 14, and the weights of nitrogen in equal volumes of that gas and of nitrogen dioxide are as 2 : 1. By using the annexed logarithmic table for the reduction of cubic centimetres of nitrogen to grams for each tenth of a degree centigrade, the calculation becomes the work of a few moments only.

Table for the reduction of Cubic Centimetres of Nitrogen to Grams.

Log $\dfrac{0 \cdot 0012562}{(1 + 0 \cdot 00367t) \, 760}$ for each tenth of a degree from 0° to 30° C.

$t°$	0.0	0.1	0.2	0.3	0.4	0.5	0.6	0.7	0.8	0.9
0°	$\bar{6} \cdot 21824$	808	793	777	761	745	729	713	697	681
1	665	649	633	617	601	586	570	554	538	522
2	507	491	475	429	443	427	412	396	380	364
3	349	333	318	302	286	270	255	239	223	208
4	192	177	161	145	130	114	098	083	067	051
5	035	020	004	*989	*973	*957	*942	*926	*911	*895
6	$\bar{6} \cdot 20879$	864	848	833	817	801	786	770	755	739
7	723	708	692	676	661	645	629	614	598	583
8	567	552	536	521	505	490	474	459	443	428
9	413	397	382	366	351	335	320	304	289	274
10	259	244	228	213	198	182	167	151	136	121
11	106	090	075	060	045	029	014	*999	*984	*969
12	$\bar{6} \cdot 19953$	938	923	907	892	877	862	846	831	816
13	800	785	770	755	740	724	709	694	679	664
14	648	633	618	603	588	573	558	543	528	513
15	497	482	467	452	437	422	407	392	377	362

Table for the reduction of Cubic Centimetres of Nitrogen to Grams—cont.

p	0·0	0·1	0·2	0·3	0·4	0·5	0·6	0·7	0.8	0·9
16	346	331	316	*301	286	271	256	241	226	211
17	196	181	166	157	136	121	106	091	076	061
18	046	031	016	001	*986	*971	*956	*941	*926	*911
19	6̄·18897	882	867	852	837	822	807	792	777	762
20	748	733	718	703	688	673	659	644	629	614
21	600	585	570	555	540	526	511	496	481	466
22	452	437	422	408	393	378	363	349	334	319
23	305	290	275	261	246	231	216	202	187	172
24	158	143	128	114	099	084	070	055	041	026
25	012	*997	*982	*968	*953	*938	*924	*909	*895	*880
26	6̄·17866	851	837	822	808	793	779	764	750	735
27	721	706	692	677	663	648	634	619	605	590
28	576	561	547	532	518	503	489	475	460	446
29	432	417	403	388	374	360	345	331	316	302

An example of the mode of calculation will serve to render the process more intelligible. Let us assume that we have made the following readings :—

	A.	B.	C.
		(After treatment with KHO)	
Volume of gas . . .	5·00 c. c.	0·40 c.c.	0·40 c. c.
Temperature . . .	15°	15·1°	15·2°
	mm.	mm.	mm.
Height of mercury in a, c, d .	300	480	480
,, ,, b . .	200	350	330
Difference	100	130	150
Add tension of aqueous vapour (Table III. Appendix) .	12·7	12·8	12·9
	112·7	142·8	162·9
Deduct for capillarity .	1·0	Add for capillarity { 2·5	2·5
	111·7	145·3	165·4
Deduct from height of bar .	760·0	760·0	760·0
	111·7	145·3	165·4
Pressure on dry gas . .	648·3	614·7	594·6

Log. of vol. of gas . .	0·69897	1·60206	1·60206
" $\dfrac{0·0012562}{(1+0·00367t)760}$.	$\bar{0}$·19497	$\bar{0}$·19482	$\bar{0}$·19467
Pressure on dry gas . .	2·81178	2·78866	2·77422
Log. of weight calc. as N. .	$\overline{3·70572}$	$\overline{\bar{4}·58554}$	$\overline{\bar{4}·57095}$

Weight calculated as N.   ·0050783   ·00038507   ·00037235

$$\text{Weight of carbon} = \frac{3(·0050783 - ·0003851)}{7} = 0·0020114$$

$$\text{Weight of nitrogen} \frac{0·00038507 + ·00037235}{2} = ·00037871$$

Sometimes, especially when the amount of carbon is large, small quantities of carbon monoxide may be formed, and may escape complete oxidation by the copper oxide placed in the anterior portion of the tube. This gas remains mixed with the nitrogen after absorption with potassium pyrogallate solution. Its amount may be determined when the whole of the gas is transferred to the measuring tube, in the last determination of volume, by removing the laboratory vessel, washing it, refilling it with mercury, and again attaching it to the face of the cap. A few drops of solution of cuprous chloride are then introduced into the vessel, and the gas allowed to act upon it. In about five minutes the absorption of the carbon monoxide will be complete ; the residual nitrogen may then be returned to the measuring tube, and its volume determined. If any carbon monoxide is found, its weight as nitrogen is calculated in the manner described, and added to that corresponding to the carbon dioxide before multiplying by $\frac{3}{7}$; its weight must also be deducted from that corresponding to the volume after treatment with potash.

Since the accuracy of this method of combustion depends upon the perfection of the vacuum obtained by the Sprengel pump, and is liable to be affected to some slight degree by nitrogen retained in the copper oxide, absorption of ammonia during the evaporation, &c., it is advisable that the experimenter should perform several blank determinations, to ascertain the accumulated effect of these errors. This should be done by evaporating to dryness in the manner described a litre of pure distilled water, with the usual quantities of sulphurous acid and ferrous chloride solutions, together with about 0·1 gram of recently ignited sodium chloride, to afford a residue which can be transferred to the tube. The residue is then to be burnt, and the gases analysed as directed: the amounts of carbon and nitrogen thus found are to be deducted from the quantities obtained in the subsequent analyses of water-residues. The corrections should amount to ·00006 gram of carbon, and ·00005 of nitrogen per litre of water. The amount of nitrogen existing as NH_3 must be subtracted from the quantity of N thus found; the remainder may be set down as organic nitrogen.

The estimation of the organic carbon and nitrogen in water is of great importance in determining the degree of organic contamination which it has experienced. Good drinking water should not contain more than 0·2 part of carbon, and 0·02 part of nitrogen per 100,000 parts of water. Sewage usually contains about four parts of carbon, and two parts of nitrogen. The ratio of carbon to nitrogen is of especial importance; the lower the ratio the more objectionable is the organic matter. The ratio in water for domestic supply may vary from five to twelve; sewage varies from one to three; polluted river-water from three to five. (For further details consult Frankland's 'Water Analysis,' Van Voorst, 1880.)

Dittmarr's and Robinson's process.—In this process the organic carbon is determined by weighing as carbon dioxide as in an ordinary combustion, whilst the organic nitrogen is converted into ammonia by ignition with soda or soda-baryta, as in the method given on p. 334.

Determination of Organic Carbon.—Evaporate 1 litre of the water, after treatment with 1 c.c. of a saturated solution of sulphurous acid, as described on p. 298, in a glass dish under the bell jar, and transfer the dried residue by the aid of a spatula to a platinum boat, which is then to be introduced into a short combustion tube, one end of which has been previously drawn out, and into which is placed (1) a spiral of silver wire gauze to reduce any nitrogen oxides which may be formed, and (2) a layer of granular copper oxide. To the drawn-out end of the tube is attached a small V-shaped tube, containing a solution of chromic acid in 60 per cent. sulphuric acid, to which is adapted a short tube filled with calcium chloride. The carbon dioxide is absorbed in a light tube containing soda-lime and calcium chloride, previously weighed. The platinum boat containing the residue having been placed in the combustion tube, the posterior end is connected by a cork and bent tube with a gas-holder containing air freed from carbon dioxide. The oxide of copper and silver are first heated, and then the platinum boat, a stream of the purified air being meanwhile sent through the apparatus. The carbon dioxide freed from water and sulphur dioxide by passing through the chromic acid solution is absorbed by the soda-lime.

Determination of Organic Nitrogen.—Place half a litre of the water in a flask connected with a condenser, and distil, collecting the distillate in the manner directed on p. 295, so as to determine the ammonia. The distillation is to be conducted until only about 30 c.c. are left in the flask. Add solutions of sulphurous acid and ferrous chloride to destroy nitrates, and if the solid residue is known to be small, add too a small quantity of potassium sulphate. Complete the evaporation under the bell jar, and when dry transfer the residue to a large silver or copper boat, moisten with a drop of water, and cover with 2 or 3 grams of a fused mixture of equal weights of baryta and soda. Introduce the boat into a short combustion tube connected at one end with a U-tube containing a known volume of water acidulated with a few drops of pure hydrochloric acid, and to the other end adapt an arrangement for sending a current of hydrogen through the apparatus. Heat the tube to redness; the organic nitrogen is converted into ammonia, which is absorbed by the dilute hydrochloric acid: its amount is estimated by the Nessler test, as described on p. 295, on one-tenth of the solution diluted to 50 c.c.

Blank experiments should be made in both determinations, and the necessary corrections introduced.

Estimation of Total Soluble Matter.—Ignite and weigh a platinum dish, place it on a glass ring on the water-bath, and fill it with the water to be examined, previously measured in a 250 c.c. flask. As the liquid evaporates, add successive portions from the flask; rinse the vessel when empty with a small quantity of distilled water, and pour the washings into the dish. When the water has entirely evaporated, heat the residue to 100°, for an hour, or until its weight is constant. The increase in the weight of the dish gives the amount of soluble matter contained in the ¼-litre of water.

*Estimation of Nitrates and Nitrites.**—This may be effected in the residue obtained from the preceding determination, by the action of precipitated copper and zinc. The apparatus seen in fig. 72 serves for the decomposition. A is a flask of 100 c.c. capacity, fitted with a caoutchouc stopper, containing (1) the tube funnel *b*, provided with a glass stopcock, and (2) the bent glass tube *c*, which is connected, by means of a well-fitting stopper, with one of the cylinders, *e*, used for the ammonia estimation (p. 292). This stopper also carries the bent tube *d*, which is about 5 millimetres in internal diameter, and is partially filled with fragments of well-washed glass. Place 3 or 4 grams of very

* Jour. Chem. Soc. June 1873.

thin sheet-zinc in small pieces in A, and cover it with a tolerably concentrated solution of copper sulphate. Allow the solution to act upon the zinc for ten or fifteen minutes, pour off the supernatant liquid, and fill up the flask several times with cold distilled water to wash the precipitated copper. After the last washing, remove the water as far as practicable. Add about 25 c.c. of distilled water to the

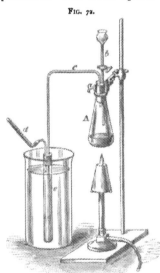

FIG. 72.

residue obtained in the determination of the total soluble matter, together with a piece of recently ignited lime, about the size of a hemp-seed, and boil the liquid (to destroy any urea which may be present), until 4 or 5 c.c. only remain. Transfer the liquid to the flask A, rinse the dish with distilled water, so as to make up the volume in A to about 15 c.c. or 20 c.c. Fit in the cork of the flask, add one drop of dilute hydrochloric acid (free from ammonium chloride) to the cylinder *e*, together with 2 or 3 c.c. of distilled water, and also moisten the glass in *d* with two drops of the acid. Fit the cork into *e*, and place the tube in a beaker of cold water, as represented in fig. 72. Heat the liquid in A to boiling, and distil it over into *e*; when A is nearly empty, fill up the funnel with hot water, turn the stopcock, allow the water to flow into the flask, and continue the ebullition until *e* contains about 40 c.c. of liquid. All the nitrates will be

reduced, and the ammonia will be expelled. Raise the
retort stand so as simultaneously to remove the tube from
the water in the beaker, and the flask A from over the lamp.
Wash the fragments of glass in *d*; the water is readily
drawn over into *e* by the contraction of the air in A on
cooling. Fill up the tube *e* to the mark, and agitate the
liquid by the aid of the bulb-stirrer. Transfer 5 c.c. to
a second tube, dilute with distilled water, add 1 c.c. of
Nessler's solution, and agitate. If the degree of colouration
is measurable, determine the quantity of ammonia required
to produce it in the manner described on p. 295; if the tint
is too dark for comparison, take a smaller quantity; if too
light (as it will be unless the water is very bad), take a
larger quantity, say 10 c.c. or 20 c.c. of the distillate, in
accordance with the indications of the preliminary trial.
The following determinations made on known quantities of
nitre may serve to show the degree of accuracy of which
this method is capable:

TAKEN. mgms.	FOUND. mgms.	TAKEN. mgms.	FOUND. mgms.	TAKEN. mgms.	FOUND. mgms.
1·67 ·	{ 1·68	2·50 ·	{ 2·42	3·34 ·	· 3·22
	1·72		2·49	4·16 ·	· 4·01

If the operator possesses the gasometric apparatus shown
on p. 303, the amount of nitrogen in the water existing
as nitrates and nitrites may be readily and accurately
estimated by determining the volume of nitric oxide
evolved on agitating the concentrated water, acidulated
with strong sulphuric acid, with metallic mercury. This
process, which is an adaptation by Frankland of Crum's
method for the refraction of nitre, is conducted as follows: The residue from 500 c.c. of the water is dissolved
in a small quantity of hot water, the chlorine precipitated
by addition of a slight excess of silver sulphate, the
liquid filtered and concentrated to a bulk not exceed-
ing 2 c.c., and is then transferred to the apparatus seen
in fig. 72A. This consists of a stout tube about 20 c.m.

long and about 1·5 c.m. internal diameter, fitted with a stopcock. The tube is filled with
mercury to the stopcock, and inverted in a basin, also containing
mercury. The concentrated filtrate is then brought into the
little cup together with the washings, and is allowed to enter
the tube by cautiously turning the
stopcock. About 1½ times the
volume of the aqueous solution
is then poured into the cup, and
thence allowed to pass into the
tube, which is then firmly closed
by the moistened thumb held
obliquely, and vigorously shaken for about five minutes.
On the completion of the reaction the gas, nitric oxide, is
transferred to the measuring apparatus, and its volume determined. If 500 c.c. of water have been used, the volume
of the nitric oxide denotes the volume of nitrogen, as nitrates
and nitrites, in 1000 c.c., since nitric oxide contains half its
volume of nitrogen. If considerable quantities of nitrites
are present, they must be oxidised to nitrates by adding a
very dilute solution of potassium permanganate to the acidulated water until the pink colour is permanent. Sodium
carbonate is then added to alkaline reaction, and the water
is evaporated and treated in the manner above described.

FIG. 72A.

Estimation of Nitrites.—A solution of meta-phenylene-
diamine ($C_6H_8N_2$) in dilute sulphuric acid gives a dark red
or reddish violet colouration with small quantities of nitrous
acid, which may be made the basis of a method for the determination of the nitrites present in natural waters. The
method requires—

    (1) *A Solution of Metaphenylenediamine.*—Dissolve 5
        grams of the base in 1 litre of water containing
        a slight excess of sulphuric acid.

(2) *Dilute Sulphuric Acid* : 1 part of acid to 2 of distilled water.

(3) *Standard Solution of pure Sodium Nitrite.*—Dissolve 0·406 gram of silver nitrite in boiling distilled water, and add a hot solution of pure common salt so long as silver chloride is precipitated. Dilute to 1 litre, allow the precipitate to settle, and make up each 100 c.c. of the clear solution to 1 litre. 1 c.c. of this solution is equivalent to 0·01 mgrm. of N_2O_3. The solutions must be kept in well-stoppered bottles, which should be quite full.

(4) Four narrow cylinders of colourless glass : these should be of such size that 100 c.c. of water rise to a height of about 18 c.: the level of the water should be marked or etched on the glass. Also several graduated pipettes or burettes.

Fill one of the cylinders to the mark with the water to be tested, and mix with 1 c.c. of the dilute sulphuric acid and 1 c.c. of the solution of meta-phenyleuediamine, and stir with the bulb tube. If a red colour appears immediately the amount of the nitrite is probably too great for accurate comparison : in that case take 50 c.c., or 25, or even 10 c.c. of the water, make up to 100 c.c., and repeat the trial. The colour should appear only at the expiration of 70 or 80 seconds.

Into the other three cylinders place measured quantities from 0·3 to 2·5 c.c. of the standard nitrite solution, dilute to the mark with distilled water, and add to each 1 c.c. of the sulphuric acid and meta-phenylene diamine solutions, and compare the tints formed with that in the first cylinder. This comparison give a first approximation. The trials are repeated on a fresh sample of the water and with the quantities of standard nitrite now deemed necessary to produce a similar tint. Lastly, a final series of comparisons, starting simultaneously, should be made, and the colours compared after the lapse of 20 minutes.

Estimation of Chlorine. By Standard Silver Nitrate and Potassium Chromate Solutions.—Dissolve 2·3944 grm. of pure dry silver nitrate in distilled water, and dilute to 1 litre; 1 c.c. of this solution is equivalent to 0·5 milligram of chlorine. The potassium chromate solution should be strong and neutral. Transfer 50 c.c. of the water to be tested to a porcelain basin, colour with 2 drops of the potassium chromate solution, and add the silver solution, drop by drop, until the permanent red colour of the silver chromate makes its appearance. If 50 c.c. have been taken, the number of c.c. of the silver solution employed gives the amount of chlorine in parts per 100,000.

In many highly-coloured waters the final point of the reaction is not distinctly visible. In such a case add a small quantity of lime-water (free from chlorides) to the measured portion of the water, pass washed carbonic acid through the liquid, boil, and filter. The colouring matter will in general be carried down by the precipitated chalk : the amount of chlorine in the filtrate may then be determined as directed above. It will be sometimes necessary, however, to destroy the organic matter by heat : the measured portion of the water, after addition of a small quantity of lime-water, must be evaporated to dryness and the residue gently heated. Treat the saline matter with hot water, filter, and determine the chlorine in the usual manner.

The quantity of chlorine contained in the water will often afford another indication of its purity. Very pure waters, as a rule, contain comparatively small quantities of chlorine, less than 1 part in 100,000 ; when contaminated with sewage (which contains on the average about 10 parts in 100,000), the quantity is largely increased. Of course in judging of the character of a drinking-water from the amount of chlorine it contains, due regard must be had to the nature of the strata through which it percolates. Water originating from springs in the neighbourhood of the sea, especially if the district be sandy, may contain considerable amounts of chlorine, and yet be free from sewage-matter.

Estimation of ' Hardness.'—Waters are familiarly spoken of as 'hard' and 'soft': these terms have reference to the action of soap upon the water. A 'hard' water necessitates the expenditure of much soap before it will give a lather ; this expenditure is caused by the action of the lime and magnesia salts, which decompose the soap, or stearate of soda, forming insoluble stearates of lime and magnesia. These substances constitute the pellicle or scum which forms upon the surface of a hard water after treatment with soap. Before a 'hard' water can give a lather useful for detergent action, sufficient soap must be used to convert all the lime and magnesia salts present into stearates. The estimation of the hardness or soap-destroying power of the water becomes therefore an important element in determining its value for economic purposes.

The soap-destroying power of the water is measured directly : a solution of soap of known strength being made to act upon a definite volume of the water until a permanent (and detergent) lather is obtained.

I. *Preparation of the Strong Soap Solution.*—Pound, in small quantities at a time, in a mortar, 3 parts of lead plaster and 1 part of dry potassium carbonate. Mix thoroughly, and add a small quantity of methylated spirit, and triturate until a thin creamy mixture is obtained. After standing for some hours, pour the clear solution through a filter, and exhaust the residue repeatedly with fresh portions of spirit. If the solution remains clear on standing, proceed to determine its exact strength by the aid of a standard solution of calcium chloride.

II. *Preparation of Standard Calcium Chloride Solution.*—Weigh out into a porcelain or platinum dish exactly ·2 gram of finely-powdered marble, cover the dish with a large watch-glass, and dissolve the marble in dilute hydrochloric acid. Heat the basin on the water-bath, and when the expulsion of the carbon dioxide is at an end, rinse the under-surface of the watch-glass into the basin and evaporate to complete dryness. Add a small quantity of water, and again evaporate to ensure the complete removal of the excess of the hydro-chloric acid. Dissolve in water, and dilute to 1,000 c.c.

III. *Dilution of the Strong Soap Solution.*—Transfer 50 c.c. of the standard calcium chloride solution to a bottle of 250 c.c. capacity, provided with a well-fitting stopper, fill up a burette with the soap solution and add it to the water in the bottle in quantities of a few drops at a time. After each addition of the soap solution insert the stopper, and shake briskly. The process is finished when a uniform lather is obtained which is permanent for at least 3 minutes, and which may be re-formed by again shaking the liquid. Read off the burette and dilute the soap solution with a mixture of 2 vols. of methylated spirit and 1 vol. of water, until about 12 c.c. of the diluted mixture are equivalent to 50 c.c. of the standard calcium chloride solution; allow it to stand for 24 hours, filter it if necessary, again determine its strength, and dilute it with the mixture of alcohol and water, until exactly 14·25 c.c. are required to produce a permanent lather with 50 c.c. of the standard calcium chloride solution.

The Process.—Transfer 50 c.c. of the water under examination to the 250 c.c. bottle, shake vigorously, and suck out the air from within the bottle by the aid of a glass tube, to remove the carbon dioxide expelled on agitating. Fill up the burette with the soap solution, and add it, 1 c.c. at a time, to the water; after each addition insert the stopper, and shake vigorously. As soon as a froth begins to form, add the solution of soap in smaller quantities until a uniform permanent lather is obtained. If more than 16 c.c. of the soap solution are required the operation must be repeated on a smaller quantity of the water. Transfer 25 c.c., or less if it is very hard, of the water to the bottle, and add sufficient distilled water to make up the volume to 50 c.c., and again add the soap solution, in small quantities at a time, until the lather is obtained. Multiply the volume in c.c. of soap solution used by the number expressing the fraction of 50 c.c. taken : thus, if 25 have been taken, multiply by 2 : if 10, multiply by 5. The weight of calcium carbonate in 100,000 parts of water, corresponding to the number of c.c. required for 50 c.c. of the water, is given in the following table : Column I. gives volume of soap solution : Column II. the corresponding amount of calcium carbonate per 100,000 parts.

Table of Hardness.

I.	II.	I.	II.	I.	II.	I.	II.	I	II.	I.	II.
c.c.		c.c.		c.c.		c.c.		c.c.		c.c.	
0·7	·00	3·3	3·64	5·9	7·29	8·5	11·05	11·1	15·00	13·7	19·13
·8	·16	·4	·77	6·0	·43	·6	·20	·2	·16	·8	·29
·9	·32	·5	·90	·1	·57	·7	·35	·3	·32	·9	·44
1·0	·48	·6	4·03	·2	·71	·8	·50	·4	·48	14·0	·60
·1	·63	·7	·16	·3	·86	·9	·65	·5	·63	·1	·76
·2	·79	·8	·29	·4	8·00	9·0	·80	·6	·79	·2	·92
·3	·95	·9	·43	·5	·14	·1	·95	·7	·95	·3	20·08
·4	1·11	4·0	·57	·6	·29	·2	12·11	·8	16·11	·4	·24
·5	·27	·1	·71	·7	·43	·3	·26	·9	·27	·5	·40
·6	·43	·2	·86	·8	·57	·4	·41	12·0	·43	·6	·56
·7	·56	·3	5·00	·9	·71	·5	·56	·1	·59	·7	·71
·8	·69	·4	·14	7·0	·86	·6	·71	·2	·75	·8	·87
·9	·82	·5	·29	·1	9·00	·7	·86	·3	·90	·9	21·03
2·0	·95	·6	·43	·2	·14	·8	13·01	·4	17·06	15·0	·19
·1	2·08	·7	·57	·3	·29	·9	·16	·5	·22	·1	·35
·2	·21	·8	·71	·4	·43	10·0	·31	·6	·38	·2	·51
·3	·34	·9	·86	·5	·57	·1	·46	·7	·54	·3	·68
·4	·47	5·0	6·00	·6	·71	·2	·61	·8	·70	·4	·85
·5	·60	·1	·14	·7	·86	·3	·76	·9	·86	·5	22·02
·6	·73	·2	·29	·8	10·00	·4	·91	13·0	18·02	·6	·18
·7	·86	·3	·43	·9	·15	·5	14·06	·1	·17	·7	·35
·8	·99	·4	·57	8·0	·30	·6	·21	·2	·33	·8	·52
·9	3·12	·5	·71	·1	·45	·7	·37	·3	·49	·9	·69
3·0	·25	·6	·86	·2	·60	·8	·52	·4	·65	16·0	·86
·1	·38	·7	7·00	·3	·75	·9	·68	·5	·81		
·2	·51	·8	·14	·4	·90	11·0	·84	·6	·97		

In waters rich in magnesian-salts the lather acquires a characteristic curdy appearance, easily recognised after a little experience. To familiarise himself with the difference in the lathers occasioned by calcareous and magnesian waters, the student should make a dilute solution of magnesium sulphate in strength equal to that of the standard calcium chloride solution, and compare the lathers obtained by adding a slight excess of soap solution to equal volumes of the two liquids. If a trial has shown the presence of magnesia salts in large proportion, the experiment should be repeated, with the water so far diluted with distilled water, that 50 c.c. of the mixture require only about 7 c.c. of the standard soap solution.

It is well known that the hardness of water is occasionally diminished by boiling: such a water contains magnesium and calcium carbonates, dissolved in free carbonic acid. On boiling the water the free carbonic acid is expelled and the carbonates are almost entirely precipitated, no diminution of the hardness will occur on boiling, unless it exceeds three parts per 100,000, since the carbonates are dissolved to that extent by water free from carbonic acid. It sometimes happens that the hardness, even when considerable, is not lessened by boiling: in this case it is due to calcium and magnesium sulphates or chlorides: such a water is termed *permanently* hard: a water which owes its soap-destroying power to carbonates is said to be *temporarily* hard. It is generally desirable to distinguish between temporary and permanent hardness in the analysis. For this purpose transfer 200 c.c. of the water to a flask, and heat to boiling. After half an hour's gentle ebullition, remove the lamp, allow the water to cool slightly, filter it through a small filter, make its volume up to 200 c.c., and again determine the hardness on 50 c.c. of the filtrate.

The number of c.c. of soap solution employed shows the permanent hardness, and this subtracted from that of the unboiled water gives the temporary hardness. Sometimes, although rarely, the hardness of the water is in part due to free hydrochloric or sulphuric acids; it is advisable, therefore, to ascertain its reaction before determining its soap-destroying power.

The hardness of water is of importance in determining its value for manufacturing purposes. Hard waters are a source of much annoyance to the manufacturer. The hard crust or 'cake' which forms in steam boilers consists of sulphate of lime and carbonates of calcium and magnesium, often mixed with co-precipitated organic matter.

Detection of Lead and Copper.—Concentrate a litre of the water to about 50 c.c., and filter it, if necessary, into one of the cylinders used for the estimation of the ammonia: add two or three drops of acetic acid, and 2 c.c. of a freshly-prepared and saturated solution of sulphuretted hydrogen. If a brown colouration is produced fill a second cylinder with distilled water, acidulate with two or three drops of acetic acid, mix with 2 c.c. of the sulphuretted hydrogen solution, and add a standard solution of lead containing $\frac{1}{10}$ milligram per c.c. (obtained by dissolving 0·1831 gram of crystallised lead acetate in a litre of distilled water) until the colouration is equal in both tubes. Copper may be accurately estimated by means of a solution of potassium ferrocyanide. The reddish-brown tint thus produced is compared with that formed under similar circumstances in distilled water containing a known quantity of copper. Iron may be determined by a similar process. (Carnelley, *Chem. News*, XXXII. 308.)

To determine if the water has any action on lead fill two beakers with the sample; into one place a bright strip of the metal, and into the other a strip which has been tarnished by previous exposure to water, and leave them in contact with the sample for 24 hours. The strips are then removed and the water tested in the manner above described. Many waters which rapidly attack a clean surface of lead have no action on the tarnished metal: other waters, especially when containing nitrates and nitrites, act on lead whether bright or tarnished.

The estimation of the amounts (1) of ammonia, (2) of organic carbon and nitrogen (or, if preferred, of the 'albuminoid ammonia '), (3) of nitrogen as nitrates and nitrites, (4) of chlorine, (5) of total soluble and suspended matter, (6) of the hardness, and (7) of the presence or absence of lead, afford the principal data in determining the value of a sample of water for domestic supply.

Occasionally it is necessary to ascertain more particularly the nature of the inorganic matter in solution : the brewer, for example, frequently wishes to know the actual amount of the calcium and magnesium sulphates and carbonates present. The estimation of the various inorganic constituents may be readily made by methods already described at length.

Estimation of Silica.—Evaporate not less than a litre of the water to dryness (best in a platinum dish), after acidifying with a few drops of hydrochloric acid. Dry the saline residue thoroughly, moisten with hydrochloric acid, dilute with hot water, and filter off the separated silica.

Estimation of Iron.—Add two drops of nitric acid to the filtrate from the silica, boil and add ammonium chloride and a slight excess of ammonia, allow the precipitate to settle, pour the supernatant liquid through a small filter, redissolve the precipitate in the least possible quantity of hydrochloric acid, and again add ammonia. Transfer the precipitate to the filter, wash with hot water, dry, and weigh the ferric oxide.

Estimation of Lime.—Add excess of ammonium oxalate to the filtrate from the preceding estimation, filter the precipitated calcium oxalate, and, after washing and drying, ignite it strongly and weigh as caustic lime.

Estimation of Magnesia.—Concentrate the filtrate from the calcium oxalate, add sodium phosphate and ammonia, and treat the magnesium-ammonium phosphate in the usual manner.

Estimation of Sulphuric Acid.—Acidify a litre of the water with a few drops of hydrochloric acid, concentrate to 80 or 100 c.c. and add excess of barium chloride solution. Filter off the barium sulphate and weigh it.

Estimation of Phosphoric Acid.—Many samples of water rich in lime and magnesia salts contain estimable quantities of this acid. To determine its amount acidify a litre of the water with nitric acid, concentrate to 50 c.c., and add solution of molybdic acid (p. 218). After standing for 24 hours treat the yellow precipitate as directed on p. 219.

Estimation of Sodium and Potassium.—Add a few drops of barium chloride to a litre of the water, to precipitate the sulphuric acid, and boil with pure milk of lime to throw down the magnesia, iron, and phosphoric acid. Filter, concentrate the filtrate, add ammonia, ammonium carbonate, and a few drops of ammonium oxalate ; again filter, and evaporate the filtrate to dryness, ignite to expel ammoniacal salts : treat the residue with a small quantity of water, filter, if necessary, acidify with hydrochloric acid, and evaporate to dryness in a weighed platinum dish. The alkalies may then be separated by platinum tetrachloride, or their proportion may be determined by dilute standard silver solution and potassium chromate.

XLI. Determination of the Amount and Nature of the Gases dissolved in Water.

Pure natural water (rain water) when thoroughly aerated contains about 20·73 c.c. of gases per litre, composed of

Nitrogen	13·08 c.c.
Oxygen.	6·37
Carbon dioxide	1·28
	20·73

The ratio of the oxygen to the nitrogen is as 1 : 2·05.

If the water becomes contaminated with putrescent organic substances the quantity of oxygen rapidly diminishes. It is abstracted from the water in oxidising the carbon, hydrogen, and nitrogen of the organic matter. By determining the ratio of the oxygen to the nitrogen in the gases which it contains, we may often ascertain whether such putrefactive changes are in actual operation in the water.

Several methods have been proposed for expelling the gases contained in water : * one of the simplest of these is due to Reichardt.† Fig. 73 represents the apparatus required for this method. A is an ordinary flask of from 800 to 1,000 c. c. Its capacity must be accurately known. The narrow cylindrical vessel B serves as a gasholder. It is fitted with a caoutchouc stopper pierced with three holes, through one of which passes the bent tube *a*, the longer limb of which ends in B at about one-third of its height

FIG. 73.

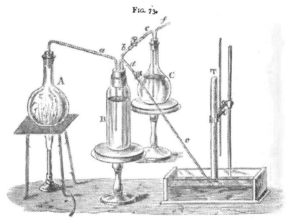

from the bottom : the other end is fitted into the caoutchouc stopper of A. The second hole in the stopper of B contains the tube *b*, which passes nearly to the bottom of the bottle : *b* is connected by means of a caoutchouc tube, which can be closed by a clamp, with the glass tube *c*, which runs nearly to the bottom of C. The third hole of the stopper of B contains the tube *d*, the end of which must be on a level with the under surface of the cork : *d* is connected with the

* See Bunsen's Gasometry, translated by Roscoe ; Miller's Inorganic Chemistry, p. 187 ; M'Leod, Chem. Soc. Jour.
† Fresenius, Zeits. für Anal. Chemie, vol. xi. p. 271.

narrow delivery tube *e*, terminating beneath the surface of
the mercury in the trough, in which is supported the absorption tube T, destined to receive the gases from the
water, and which is therefore filled with mercury before the
commencement of the experiment.

The flask A is completely filled with the water to be
examined ; B and C are also partially filled with recently
well-boiled and still warm distilled water, and *b* and *c*, *d* and
e, are connected together ; both the clamps being open. By
simply blowing through *f*, the bottle B and all the tubes are
completely filled with the warm water : the clamps of *b* and
d are now successively closed, and the tube *a* is inserted
into the flask A, care of course being taken that no air-
bubbles lodge beneath the surface of the cork. Open the
clamp of *b* and *c*, and gradually boil the water in A. The
gas expelled on ebullition collects in B : as soon as the
whole is eliminated, which will require at least an hour, and
the water in B is nearly boiling, open the clamp of *d* and *e*,
cautiously blow through *f*, so as to expel the water contained in *d* and *e*, which is allowed to flow out into the trough,
and displace the gases from B into the measuring tube by
again blowing through *f*. On removing the lamp from beneath
A, the water, on receding, should completely fill the flask. If
it does not, continue the ebullition and collect the small
portion of the remaining gas.

Allow the *moist* gas in the tube T to acquire the temperature of the air (obtained from the thermometer *t*). Read off
the level of the mercury within and without the tube,
together with the temperature and barometric pressure at
the time of observation, and calculate the volume of the
gas *when dry* at 0° and 1 metre pressure. Introduce a ball
of potash attached to a platinum wire into the tube, and in
six or eight hours again read off the levels of the mercury.
Quickly withdraw the piece of potash, moisten it slightly
with water, and reintroduce it into the tube, and at the
expiration of another hour again determine the levels of the

mercury, the temperature, and atmospheric pressure. If the second and third readings are identical the absorption of the *carbon dioxide* is complete. Reduce the volume to o° and 1 metre pressure: it is not necessary to deduct the tension of aqueous vapour corresponding to the temperature observed, since the gas may be assumed to be dry after having been in contact with the potash. Bring into the tube a papier-mâché bullet, soaked with a solution of potassium pyrogallate, and from time to time read off the level of the mercury within the tube. As soon as the contraction appears to be finished withdraw the papier-mâché bullet, and accurately determine the position of the levels of the mercury within and without the tube, together with the temperature and pressure. Reduce the volume of the residual gas (*nitrogen*) to o°C and 1 metre pressure. The volume absorbed by the alkaline pyrogallate represents the amount of *oxygen*.

The potash balls for the absorption of carbon dioxide may be readily made by inserting the end of a platinum wire into a little notch filed in the face of a bullet-mould, on the opposite side to the orifice through which in bullet-making the fused lead is poured. A small quantity of potash is melted in a silver dish and poured through the hole until the mould is completely filled. When quite cold the potash ball may be readily detached from the metal. The short projecting piece formed by the hole through which the potash has been poured should be cut away and the ball preserved in a stoppered bottle until used.

The papier-mâché balls are made in a similar way: filter-paper, converted into pulp by maceration in water, is forced through the hole into the mould until it is quite filled, and the mould is placed in the steam chamber and heated until the paper is dry.

Of course if the operator possesses the apparatus described on p. 303 *et seq.*, the accurate analysis of the gas by the use of liquid reagents becomes the work of a few minutes only.

MANURES.

XLII. GUANO.

THIS substance is the excrement of sea-birds, more or less altered by exposure to the weather. It contains ammonia in combination with uric, oxalic, carbonic, and phosphoric acids, phosphates and sulphates of lime, magnesia, and alkalies, and more or less organic matter, water, sand, &c. It is very variable in composition, and is often largely adulterated. Its fertilising power mainly depends upon the ammonia and phosphoric acid which it contains.

Mix the sample carefully, and transfer about 50 grams to a stoppered bottle ; the several portions used in the analysis are to be taken from this quantity.

1. *Determination of the Moisture.*

Weigh out about 5 or 6 grams of the guano into the tube, fig. 29, p. 70, heat the oil-bath to 120°, aspirate a slow current of dry air through the apparatus, and repeatedly weigh the tube until the weight is constant. The flask contains 5 c.c. of normal acid, diluted with water, to absorb the ammonia, which volatilises with the steam when guano is heated ; the quantity of residual acid may be determined with litmus and a dilute soda solution in the usual way. The loss in the weight of the tube, *minus* the amount of ammonia retained in the flask, gives the quantity of moisture.

2. *Determination of Fixed Inorganic Matter.*

About four-fifths of the dried guano is transferred from the tube to a weighed platinum dish. Re-cork the tube securely ; the remaining portion of the dried guano will be used for the estimation of the nitrogen. Gently ignite the portion in the dish. The ash should be nearly white ; if it is of a reddish colour, adulteration with sand or clay may be suspected. The quantity of the ash in the better class of guanos does not exceed 35 per cent. The loss of weight gives the organic matter, together with the ammonia and combined water.

3. *Determination of the Insoluble Matter, Sand, &c.*

Boil the weighed portion of the ash with dilute nitric acid for 15 minutes.* If the substance effervesces strongly the guano has in all probability been adulterated with calcium carbonate. Heat on the water-bath for some time, add water, filter into a 300 c.c. flask, wash the residue, dry, and weigh it.

4. *Determination of the Phosphoric Acid, Lime, Magnesia, Sulphuric Acid, and Alkalies.*

Make up the filtrate to the containing-mark, and shake the liquid.

(*a*) *Phosphoric Acid, Lime, and Magnesia.* — Transfer 100 c.c. of the solution to a ½-litre flask, add excess of ammonia, and then acetic acid to acid reaction. Dilute the liquid to the mark, and shake.

a. Phosphoric Acid.—By Standard Uranium Solution.— When a solution of acetate or nitrate of uranium is added to a solution of phosphoric acid, containing ammoniacal salts and free acetic acid, a light greenish-yellow precipitate of the double uranium-ammonium phosphate is produced. This precipitate is insoluble in water, and in acetic acid ; the stronger acids, however, dissolve it.

A solution of the acetate or nitrate of uranium gives a reddish-brown colour with potassium ferrocyanide. This colour is not produced so long as any phosphoric acid is in solution. These reactions form the basis of an accurate volumetric method for the estimation of phosphoric acid.

Preparation of the Standard Solution of Uranium.— Dissolve about 35 grams of well-crystallised uranium nitrate or acetate (the former, however, is preferable) in 900 c.c. of water. The solution, mixed with sodium acetate, must be standardised by a solution of pure sodium

* In burning off the organic matter the phosphates are partially converted into pyrophosphates. By boiling with dilute nitric acid the pyrophosphates are reconverted into orthophosphates.

phosphate of known strength. Dissolve 100 grams of sodium acetate in 900 c.c. of water, and dilute with strong acetic acid to 1 litre. Dissolve 10·085 grams of pure and non-effloresced crystals of sodium phosphate, previously dried by pressure between filter-paper, in a litre of water.

Transfer 50 c.c. of the sodium phosphate solution, corresponding to 0·1 gram P_2O_5, to a beaker, add 5 c.c. of the sodium acetate solution, and heat the mixture on the water-bath. Remove it when its temperature is about 80°, and quickly run in 10 or 12 c.c. of the uranium solution, with constant stirring. Now add the uranium solution more cautiously, in quantities of 0·5 c.c. at a time, and test the mixture after each addition. For this purpose bring a few drops of the turbid, but nearly colourless, liquid, on to a porcelain slab, and add to it a small drop of potassium ferrocyanide solution; if the least excess of uranium salt be present, the mixed drops will acquire a reddish-brown colour. If this colour is not at once perceived, continue the addition of the uranium solution until it just appears. Replace the beaker on the water-bath, and in a few minutes again transfer a few drops to the slab, and test a second time with the ferrocyanide. If the colouration is still distinctly visible, the process is finished. The solution of the uranium salt must now be diluted, until 20 c.c. are exactly required for the 50 c.c. of sodium phosphate solution : 20 c.c. thus become equivalent to 0·1 gram of P_2O_5, or 1 c.c. = 5 milligrams P_2O_5. The exact strength should be again determined by several trials after the dilution has been effected.

To apply this process to the determination of the phosphoric acid contained in the solution of the guano, transfer 50 c.c. of the acetic acid solution (*a*) of the phosphate to a beaker, heat on the water-bath, and proceed exactly as described. Repeat the determination on a second portion of the liquid.

β. Lime.—Transfer 100 c.c. of the liquid (*a*) to a beaker, heat, and add excess of ammonium oxalate; the calcium oxalate is filtered off after standing, and weighed as carbonate, or it is treated with sulphuric acid, and titrated with potassium permanganate.

γ. *Magnesia.* The filtrate from the lime precipitate is mixed with ammonia and sodium phosphate, and the magnesium-ammonium phosphate weighed as pyrophosphate.

(*b*) *Determination of Sulphuric Acid and Alkalies.*— Transfer the remainder (200 c.c.) of the liquid in 4 to a beaker, heat, add barium chloride, and filter off any barium sulphate which may form. To the filtrate add milk of lime or baryta water, boil, filter, and precipitate the excess of the alkaline earth with ammonia and ammonium carbonate, filter, evaporate to dryness, and ignite, to expel the ammoniacal salts. The residue is treated with a small quantity of water, filtered again, if necessary, and the filtrate evaporated to dryness in a weighed platinum dish. The proportion of the alkalies in the washed chlorides may then be determined by a dilute standard solution of silver nitrate and potassium chromate.

5. *Determination of the Nitrogen.*

α. *Nitrogen existing as Ammonia.*—From 1 to 5 grams of the guano, according to its supposed richness in ammonia, as determined by the amount evolved in drying, are weighed out into the retort, fig. 32, and boiled with magnesia for some time. The ammonia is gradually expelled, and may be collected in the flask c, containing a known quantity of standard sulphuric or hydrochloric acid, diluted with water. The quantity of the residual free acid is then to be estimated by standard soda and litmus solution. The use of caustic soda or lime in expelling the ammonia is inadmissible, since these substances would convert a portion of the nitrogenous organic matter into ammonia.

β. *Nitrogen existing as Azotised Organic Matter, Uric Acid, &c.—By Ignition with Soda-lime.*—Many organic substances containing nitrogen, not in the form of nitroxides, when heated with a caustic alkali, give up the whole of their nitrogen in the form of ammonia. This reaction constitutes the principle of a convenient method for estimating the amount of organic nitrogen existing in manures.

The following articles are required for this method :—

(*a*) Combustion-tube.—This should have the form seen in fig. 74 ; it is about 40 centimetres long, and from 10 to 12 millimetres in internal diameter.

(*b*) Soda-lime.—Heat a sufficient quantity of the coarsely powdered substance in a porcelain basin, just before it is wanted, and allow it to cool. A mixture of equal weights of dry slaked lime and dehydrated sodium carbonate may be used instead of soda-lime.

(*c*) Oxalic Acid.—This should be well dried in the water-bath so as to expel all its water of crystallisation.

(*d*) Asbestos.—Ignite a small quantity in the gas-flame before use.

(*e*) A bulbed U-tube, fitted with caoutchouc stopper and bent tube. On the end of the bent tube is a cork, which fits tightly into the combustion-tube.

Introduce a layer, about 3 centimetres in length, of the dried oxalic acid, mixed with a small quantity of soda-lime, into the posterior end of the tube, and afterwards an equal bulk of soda-lime. Weigh out from the tube the remainder of the dried guano obtained in 1, into a dry porcelain mortar, and mix it with soda-lime. Bring the mixture without loss of time (since it is apt to part with a small quantity of ammonia) into the tube, and rinse out the mortar with a fresh portion of soda-lime. The substance should be mixed with a sufficient amount of soda-lime to occupy about 20 centimetres of the length of the tube. Fill up the tube to within 5 centimetres of its length with soda-lime, and insert a loosely-fitting plug of the recently-ignited asbestos. Fit in the cork of the U-tube, transfer 10 c.c. of standard acid to the U-tube, and add sufficient water to fill the bulbs to the extent indicated in the figure. Gently tap the combustion-tube on the table

so as to make a passage for the evolved gases. Place
the combustion-tube in the furnace, and gradually heat it
along its entire length, beginning at the end nearest the
U-tube. The heat must be sufficient to cause a steady
evolution of gas; towards the end it should be increased, to
break up any cyanides which may have been formed. Do
not heat the extreme end of the tube where the oxalic acid
is situated, until the evolution of the gas has almost
finished. When the combustion is at end, and the evolution
of gas has *totally ceased*, cautiously heat the oxalic acid ;
this occasions a brisk current of carbon dioxide, which
sweeps out all the ammonia remaining in the tube. Remove
the U-tube when the evolution of the gas has nearly finished,

FIG. 74.

add a few drops of litmus solution, and dilute caustic
soda solution from a burette, until the free acid is nearly
neutralised. Transfer the liquid to a beaker, wash out the
bulbs, and complete the addition of the soda solution. By
operating in this manner there is less chance of loss arising
from incomplete transference of the acid liquid, and less
washing water is afterwards required. It will be found
most convenient to use a soda solution, of which about 3 c.c.
are equivalent to 1 c.c. of normal acid. If it is preferred to
determine the nitrogen by weight (and this should be done
if much empyreumatic matter be present in the liquid of the
bulbs), rinse the bulbs into a beaker, and pour the solution
through a moistened filter. Evaporate the filtrate to dryness
with excess of platinum tetrachloride, transfer the washed
double salt to a weighed porcelain crucible, in the manner
directed on p. 84, and dry it slowly ; heat the crucible

to bright redness, and weigh the residual platinum. 194·4 parts of platinum are equivalent to 28 of nitrogen. The amount of the nitrogen cannot be calculated from the weight of the double salt, since it is apt to contain considerable quantities of compounds of platinum with organic bases. These bases, however, contain the same proportion of nitrogen and platinum as the ammonium-platinum chloride. By determining the amount of platinum left on ignition, the proportion of the nitrogen is therefore readily calculated.

Deduct the amount of nitrogen corresponding to the ammonia found in *a* : the difference shows the quantity of organic nitrogen.

XLIII. Bone-Dust.

1. *Moisture.*—Dry a weighed portion at 120-130° in the air-bath.

2. *Carbonic Acid.*—Determine this constituent by means of the apparatus represented in fig. 31, p. 86.

3. *Fixed Constituents.*—See p. 331.

4. *Insoluble Matter and Sand, &c.*—See p. 332.

5. *Soluble Matters after Treatment with Hydrochloric Acid.* —See p. 332.

6. *Fat.*—Treat about 6 or 8 grams of the sample with boiling ether in an apparatus adjusted for distillation *per ascensum*, and dry the insoluble matter at 120-130° in the air-bath. From the loss of weight deduct the moisture found in 1 ; the remainder gives the quantity of fatty matter.

7. *Gelatigenous Matter.*—Add together the amounts of the several constituents : the difference required to make up 100 may be set down as gelatigenous substance.

XLIV. Superphosphates.

The phosphoric acid existing in the majority of naturally occurring phosphates is not very readily dissolved by water, and is therefore not in the form in which it can be rapidly assimilated by plants. The manure manufacturer converts a portion of the phosphoric acid into the soluble modification by treating the phosphorite, bone-dust, spent bone-black, &c., with sulphuric acid. In this operation the insoluble tricalcium phosphate ($Ca_3P_2O_8$) is converted into the soluble monocalcium phosphate, $CaH_4P_2O_8$, calcium sulphate being simultaneously produced. The pasty mass which runs out of the apparatus in which the mixture of phosphate and acid is made, gradually becomes dry on standing, partly from the evaporation, and partly from the assimilation of the water. To increase the fertilising power of the material, or to satisfy the tastes of the consumers, the manufacturer frequently adds various substances, such as dried or liquid blood to the material before or after treatment with the acid. Superphosphate therefore consists essentially of monocalcium phosphate, $CaH_4P_2O_8$ (so-called soluble phosphate), mixed with tricalcium phosphate, $Ca_3P_2O_8$ (insoluble phosphate), calcium sulphate, oxides of iron, alumina, magnesia, and alkalies, and more or less organic matter and moisture.

Sample the mixture carefully, and transfer a portion to a stoppered bottle, from which the quantities employed for the several estimations are to be taken.

1. *Water.*—Dry a weighed portion of the sample at 170° in the air-bath until it ceases to lose weight. The loss gives the quantity of moisture and water existing in combination with the calcium sulphate.

2. Weigh out 10 grams of the undried superphosphate into a mortar, add a small quantity of cold water, and triturate with the aid of the pestle; allow the suspended matter to settle for a few minutes, and pour the liquid through a filter into a 500 c.c. flask. Repeat the extraction with cold water in the same manner several times in succession, and finally wash the residue with hot water, transferring the washings to the filter. Dilute the filtrate to the mark and shake the liquid. Weigh the insoluble portion.

I. *Examination of the Filtrate.*

(*a*) *Estimation of the Ferric Phosphate and Soluble Calcium Phosphate.*—Transfer 200 c.c. of the liquid to a platinum dish, and evaporate, adding an excess of sodium carbonate and a little nitre so soon as the whole of the liquid has been brought into the dish. When the mass is dry, ignite gently, mix with a little water, and rinse the contents of the dish into a beaker, add excess of hydrochloric acid, and heat until the liquid is clear. Mix with excess of ammonia, acidulate with acetic acid, and filter off and weigh the *ferric phosphate*, receiving the filtrate in a 250 c.c. flask. Dilute to the mark and shake. Withdraw successive portions of 50 c.c., and determine the *phosphoric acid* by uranium solution. Transfer 100 c.c. of the liquid to a beaker, and determine the *lime* and *magnesia*, as directed on p. 333.

(*b*) *Estimation of Organic Matter and Alkalies.*—Evaporate 100 c.c. in a platinum dish, adding milk of lime until the liquid is distinctly alkaline. Dry the residue at 180° and weigh. Ignite the dried mass and again weigh; the difference gives the quantity of *organic matter*. Boil the weighed residue with lime water, then with pure water; filter, and add barium chloride to precipitate the sulphuric acid; mix with ammonia, ammonium carbonate, and oxalate; filter, evaporate to dryness with hydrochloric acid, ignite the residue, treat with water, filter, and weigh the *alkaline chlorides*.

(*c*) *Determination of the Sulphuric Acid.*—Heat 100 c.c. of the liquid in a beaker, acidulate with a few drops of hydrochloric acid, and precipitate with barium chloride.

II. *Examination of the Insoluble Portion.*

(*a*) *Determination of the Carbon.*—Ignite gently in a platinum dish; the loss of weight gives the quantity of *organic matter and charcoal*.

(*b*) *Determination of Sand, Clay, &c.*—Boil the ignited portion repeatedly with dilute hydrochloric acid, filter into a ½-litre flask, and wash with hot water. The insoluble residue consists of sand and clay. Dilute the filtrate to the mark and shake.

(*c*) *Determination of Phosphoric Acid, Iron, Lime, and Magnesia.*—Transfer 100 c.c. of the above solution to a beaker and proceed exactly as in I. (*a*).

(*d*) *Determination of Sulphuric Acid.*—In 100 c.c., by barium chloride in the usual manner.

(*e*) *Determination of Total Nitrogen.*—In from 1 to 2 grams of the original substance by ignition with soda-lime. (See p. 335.)

(*f*) *Determination of Ammonia.* — Superphosphates are occasionally mixed with ammoniacal salts. The amount of this ammonia is determined as in No. VIII. Part II.

The results of the analysis should be arranged according to the subjoined form :

Soluble constituents ⎰ Phosphoric acid*
⎱ Lime
⎱ Magnesia
⎱ Ferric oxide

Insoluble constituents ⎰ Phosphoric acid†
⎱ Lime
⎱ Magnesia
⎱ Ferric oxide
⎱ Alumina

Total calcium sulphate, .
„ organic matter and charcoal,‡ .
„ sand and clay, .
„ moisture, .

* Equal to   per cent. soluble phosphate.
† Equal to   per cent. insoluble phosphate.
‡ Containing   per cent. of nitrogen, equal to   per cent. ammonia.

XLV. Ashes of Plants.

The substances generally present in estimable quantity in the ashes of plants are silica, phosphoric, sulphuric, and carbonic acids, chlorine, potash, soda, lime, magnesia, iron, and manganese.   In much smaller quantity are sometimes found alumina, lithia, strontia, baryta, rubidia, copper, fluorine, iodine and bromine, cyanides and cyanates, boracic acid, sulphides, &c.

Certain of these substances are, however, never present in plants: thus the cyanides and cyanates are formed by the mutual action of the carbon and nitrogen in the plant at the high temperature of the incineration: in the case of ashes rich in alkalies and alkaline earths they may have been also formed by the action of the nitrogen in the air during the burning. Probably too, all the sulphur found in the ash did not originally exist as sulphuric acid: not unfrequently sulphur exists in the unoxidised state in a plant, and in combination with carbon, hydrogen, and nitrogen, forming peculiar organic acids. In presence of the bases the sulphur becomes converted into sulphuric acid during the incineration: sometimes a portion of the sulphur escapes oxidation, or when oxidised is again reduced by the admixed charcoal, giving rise to the sulphides. The main quantity of the carbonic acid present in the ash is derived from the destruction of organic acids combined with the alkalies.

In order to obtain the ash in a proper state for analysis, the portions of the plant to be incinerated must be freed as far as possible from adhering soil, &c., by brushing or rubbing. In the case of small seeds the best plan is to treat them in a beaker with a small quantity of water, stir them with a glass rod for a minute or so, and throw them on a sieve, the meshes of which are sufficiently coarse to allow the sand to pass through, whilst retaining the seeds. Repeat this operation several times, but take care not to allow the seeds to remain too long in contact with the water, or portions of the soluble salts will be dissolved out. Place the seeds in a cloth and rub them between its folds, and dry them on a water-bath. The substance to be incinerated is weighed, and placed in a shallow porcelain basin fitting into a muffle, which is to be gradually heated to low redness.

Great care must be taken duly to regulate the heat: if it is too high, the process of incineration will be retarded: the salts will fuse, and enclose the carbonaceous matter, thus protecting it from the action of the air.   Moreover, at a high temperature, chloride of sodium would volatilise, and a part of the phosphorus would be lost.   It seldom facilitates the operation to stir the heated mass, as its porosity and looseness of aggregation are thereby destroyed. The supply of air must be adequate, but not excessive, otherwise particles of the ash are apt to be carried away in the draught.   In the

ash of vegetables, the amount of alkali is frequently so considerable that it is almost impossible to obtain the mass quite white at a temperature sufficiently low to prevent it fusing.   In this case it is best to char the body in a Hessian crucible at a low red heat (scarcely visible in daylight), extract the soluble portion with water, and complete the incineration of the residue in a muffle. In all cases the ash must be weighed, properly mixed in a smooth porcelain crucible, and preserved in a well-stoppered bottle.

FIG. 75.

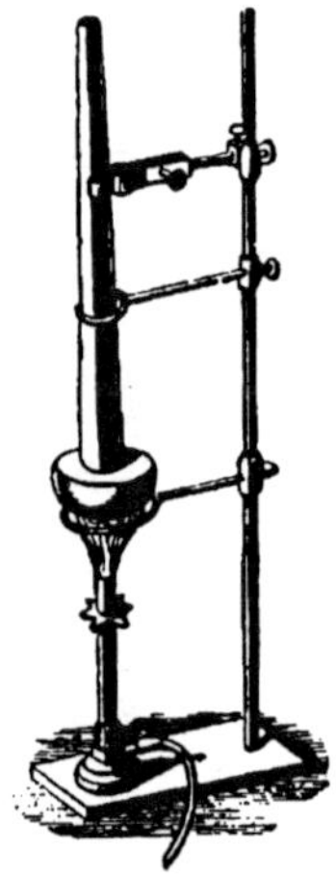

The ash of organic substances may in general be readily obtained free from carbonaceous matter by the simple arrangement seen in fig. 75.   The mass is charred in a porcelain dish at a low red heat, and as soon as the evolution of empyreumatic matter ceases, the neck of a large retort is supported over the dish by means of a clamp, and the heating is continued until the mass is white. The increased current of air playing over the heated mass facilitates the combustion of the carbon.   This method is liable, however, to increase the amount of sulphates present in the ash, owing to the action of the sulphuric acid derived from the coal gas : in cases where great accuracy is required the Bunsen lamp must be replaced by a spirit lamp.

From 7 to 10 grams of the well-mixed and finely-powdered ash are placed in a glass cylinder of about 300 cubic centimetres capacity, provided with a well-fitting stopper. About 25 cubic centimetres of distilled water are then added, and carbon dioxide is passed into the cylinder. The delivery tube of the apparatus (which must not dip into the liquid) is occasionally withdrawn, the stopper inserted, and the liquid shaken to promote the absorption of the gas. When the caustic bases are completely neutralised and the solution saturated (which is evidenced by the cessation of the partial vacuum, and also by the bubbles passing *upwards* between the bottle and its stopper when the latter is cautiously lifted after the liquid has been shaken), the contents of the cylinder are washed into a porcelain dish,* evaporated to complete dryness, again heated with a small quantity of water to dissolve the alkaline salts, and after standing a short time

Fig. 76.

filtered through a weighed filter. The filtrate is again evaporated to dryness, the saline residue treated with a small quantity of water, and the calcium sulphate which separates out filtered off through a weighed filter. The filtrate is received in a small weighed flask of 150 cubic centimetres capacity, provided with a side tubulus (fig. 76). This is easily made by directing the flame of the blowpipe upon the side of the flask until the glass is softened, when on touching the softened part by a thick platinum wire it will adhere, and a portion of the glass may be drawn out in the form of a narrow tube. The wire is detached from the tubulus by scratching the latter with a cutting diamond. Care must be

* If calcium carbonate crystallises on the side of the cylinder it may be removed by adding a little water, saturating it with carbonic acid, and dissolving the thin crust by vigorously shaking the liquid.

taken in filtering the liquid containing the soluble portion of the ash, into this flask that the end of the funnel does not dip into the liquid ; the funnel must be maintained in such a position that the drops in falling into the flask are not splashed against its upper sides. The filtrate is diluted to about 60 cubic centimetres, and well mixed by shaking. The edge of the tubulus is slightly greased, and the flask and solution weighed. The liquid is then divided into six portions, contained in little beakers, to serve for the determination of the sulphuric acid, alkalies, chlorine, phosphoric and carbonic acids, the sixth portion being reserved in case of accident. The object of the tubulus is to allow of the liquid being poured from the flask into the beakers : the amount taken for each determination is indicated by the loss of weight suffered by the flask and solution.

The carbonic acid is determined volumetrically by deci-normal sulphuric acid and litmus solutions ; the sulphuric acid and chlorine by precipitation as barium sulphate and silver chloride. The portion for the phosphoric acid determination is acidified with hydrochloric acid solution, boiled to expel carbonic acid, allowed to cool, ammonia added, together with a few drops of magnesia-mixture, and the magnesium-ammonium phosphate weighed as pyrophosphate. In order to determine the amount of the alkalies, the solution is boiled with a slight excess of baryta water (best in a platinum or silver dish); the sulphuric, carbonic, and phosphoric acids, together with the greater portion of the magnesia dissolved, are thus separated : the excess of baryta is removed by ammonia and ammonium carbonate. The filtrate is evaporated to dryness in a platinum dish, gently heated, re-dissolved in a few drops of water, filtered if necessary, a few drops of hydrochloric acid added, and the liquid evaporated to dryness, heated, and the mixed alkaline chlorides weighed. The potassium chloride is then separated by platinum tetrachloride, or the relative amount of the two chlorides determined by standard silver. In cases where the amount of the

soluble portion of the ash is comparatively large, more than
traces of magnesia will remain in solution with the alkaline
salts. This portion of the magnesia is found in the filtrate from
the double chloride of potassium and platinum : its amount
may be estimated by evaporating the alcoholic solution to
dryness, re-dissolving in water, and
transferring the liquid to a small flask
provided with a tightly-fitting cork,
furnished with two tubes, as in fig.
77. Hydrogen is led through the
tube a, and the exit tube b, within
the flask, is sufficiently long to reach
just above the surface of the liquid,
so as to ensure the thorough expulsion of the air. When the flask is
completely filled with hydrogen, the
ends of the tubes are closed by
stoppers, and the flask is placed in direct sunlight, when the
platinum is quickly reduced to the metallic state, and the
solution becomes colourless. The process of reduction
may, if necessary, be facilitated by heating the solution on a
water-bath before the transmission of the gas. If the capacity of the flask is small, it will be requisite to refill it once or
twice with hydrogen to ensure the complete reduction of the
platinum ; it is then desirable to displace the remaining gas
by a rapid current of carbonic acid, otherwise an explosion
might occur, particularly if the contents of the flask are
warm, owing to the surface action of the finely-divided
platinum on a mixture of air and hydrogen. The colourless
solution is then filtered from the reduced metal, and, after
concentration, the magnesia precipitated by sodium phosphate and ammonia. This method is recommended to be
used in all accurate separations of the alkalies from magnesia :
it is moreover a rapid and easy mode of recovering the excess
of platinum used in the determination of potassium or ammonium salts.

FIG. 77.

In the insoluble portion are contained lime, magnesia, ferric oxide (alumina), silica, phosphoric, sulphuric, and carbonic acids. This is dried at 100° and weighed. It is detached as far as possible from the filter, and the latter incinerated. The ash from the filter-paper is allowed to fall into a porcelain basin, and treated with water saturated with carbonic acid, evaporated *to perfect dryness* on the water-bath, and mixed with the main quantity of the insoluble portion in a smooth porcelain mortar. The carbonic acid is determined in about 1 to 2 grams of the substance according to the method given in No. V. Part II. ; the solution in the flask serves for the determination of the silica, sand, charcoal, and sulphuric acid. The phosphoric acid, iron (alumina), manganese, lime, and magnesia are determined in about 2 grams of the remainder of the insoluble matter. The weighed portion is dissolved in nitric acid, and after separation of the silica in the usual manner, the solution is again evaporated nearly to dryness in a porcelain basin, and dilute nitric acid added until the bases are completely dissolved, strong fuming nitric acid (saturated with the lower oxides of nitrogen) added until calcium nitrate begins to separate : a few more drops of dilute nitric acid are now added to destroy the slight turbidity. The nitric acid solution of the substances is thus in the highest possible state of concentration. It is covered with a large watch-glass, gently warmed, and about 2 grams of tin-foil added in small portions at a time. The tin is rapidly oxidised, and the supernatant liquid becomes perfectly clear. The preliminary heating of the solution is absolutely necessary, since in the cold the metal is apt to become passive, when it resists the action of the acid. Care must be taken to keep the nitric acid in sufficient excess, in order to prevent the formation of hydrated monoxide, which renders the solution inconveniently turbid. When all action is at an end, and the tin fully oxidised, the contents of the dish are evaporated *nearly* to dryness, water is added, and the solution filtered. The precipitate contains

all the phosphoric acid ; the bases are found in the filtrate.
The precipitate, detached as far as possible from the filter, is
digested in the smallest possible quantity of highly-concen-
trated potash solution ; on the addition of water the solution
will become perfectly clear, provided no great excess of the
alkali has been used. The small amount of the precipitate
still adhering to the filter is also dissolved in a few drops of
potash solution, and added to the main portion of the liquid.
The mixture is then saturated with sulphuretted hydrogen,
acetic or sulphuric acid added in very slight excess, and the
precipitated tin sulphide separated by the filter-pump. The
filtrate is concentrated to a small bulk, filtered from the
slight amount of tin sulphide, which often separates on
evaporation, and the phosphoric acid precipitated by magnesia-mixture and ammonia. The filtrate from the insoluble
tin phosphate is treated with sulphuretted hydrogen to
remove the lead with which the foil is frequently mixed,
filtered, evaporated to a small bulk, boiled, ammonia added
in slight excess, and the iron and alumina filtered off : they
are separated as in No. XII. Part II. The filtrate from the
precipitate by ammonia contains the manganese, lime, and
magnesia. These are separated as in No. XIX. Part IV.,
p. 220.

PART V.

ORGANIC ANALYSIS.

I. ANALYSIS OF BODIES CONTAINING CARBON AND HY-
DROGEN, OR CARBON, HYDROGEN, AND OXYGEN.

ORGANIC substances containing hydrogen, when heated
with cupric oxide, are converted into carbon dioxide and
water. By absorbing the products of the combustion in a
suitably-arranged apparatus, and weighing them, we can
readily calculate the amount of carbon and hydrogen in
the substance analysed, from the knowledge that 44 parts
of carbon dioxide contain 12 parts of carbon, and that
18 parts of water contain 2 parts of hydrogen. If the sum
of the amounts of carbon and hydrogen is equal to the
weight of the body taken, the substance contains only these
elements ; if the body contains oxygen in addition, the difference indicates the amount of this constituent.

Fig. 78 represents the apparatus in which the combustion
may be conveniently made. The substance is burnt with
cupric oxide by the aid of a current of oxygen or air. The
gas-furnace is of the form known as Erlenmeyer's ; it consists
of 24 Bunsen-burners, each provided with a separate stopcock worked by a little lever. The width of the air-passages
in the burners may be regulated by a short piece of moveable tube, so that the amount of air passing into the tube
may be altered at will. This arrangement serves to prevent
the flame passing down to the burner at the bottom when
the gas-current is feeble. The tubes end in a horizontal pipe,
which is connected with the gas-supply by wide caoutchouc
tubes. The flames strike against a semi-circular trough of
well-baked fire-clay, resting on small clay supports ; in this
trough is placed the combustion-tube. The side plates *a, a*

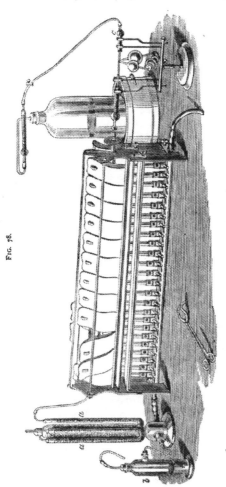

FIG. 78.

(fig. 79) are of clay ; they are moveable, and are supported upon a ledge running the entire length of the furnace.

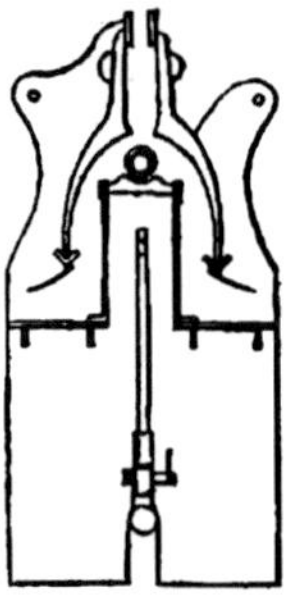

It will be seen from their peculiar shape (fig. 79) that the flames, after diverging from beneath the trough, strike against the sides ; the heat is thus reverberated, and the tube is uniformly and regularly heated. By the aid of the clamping screws a slight inclination may be given to the ledge on which the plates rest, or. it may be raised or lowered above the burners.

FIG. 79.

The following articles are needed to make a combustion by means of this apparatus :—

1. *A Piece of Combustion-tube.* This should be about 4 or 5 centimetres longer than the furnace ; and it should be about 2 millimetres thick in the glass, and about 12 or 14 millimetres in internal diameter. The sharp edges of the tube should be fused in the blowpipe flame, so that two caoutchouc stoppers, pierced with holes, may be introduced without being cut or torn.

2. *A Calcium Chloride Tube.* This serves to absorb the water produced : it may conveniently be arranged as in fig. 80. It is furnished with two bulbs, *a* and *b* : in the small neck between the bulbs is fused a piece of thin glass tube projecting into the bulb *a*. By carefully regulating the heat, the greater portion of the water produced in the combustion condenses in *a* : if its quantity is not too considerable, it remains in this bulb when the tube is held perpendicularly with the bulbs uppermost. After having been weighed the water may readily be emptied out into a little capsule, and its purity tested by its taste, smell, action on litmus-paper, &c. A calcium chloride tube so arranged may be used for a great

FIG. 80.

number of observations without replenishing, provided that on the conclusion of the experiment the bulb be emptied and the tube dried by the aid of a narrow roll of filter-paper. To fill the calcium chloride tube, place a loose plug of cotton-wool within the wide tube, close the end with the finger, and suck out the air at the narrow end. On suddenly removing the finger, the loose plug is driven into the larger bulb : repeat this operation until the long fibres of the wool are within the neck between *a* and *b* : these fibres tend to prevent the formation of drops in the narrow tube, and thus to promote the regularity of the· passage of the gas through the potash bulbs. Fill the larger bulb with coarse fragments of spongy calcium chloride, gently tapping the tube so as to shake the pieces together, and then add smaller pieces (not powder) until the tube is nearly filled ; insert a plug of cotton-wool and close the tube with a good, softened, tightly-fitting cork, through which passes a tube about 4 centimetres long and of the same diameter as the tube of the potash bulbs. Fuse the sharp edges of the tube before inserting it into the cork. After fitting the cork into the calcium chloride tube, cut the protruding portion with a sharp knife in the manner seen in fig. 80, and neatly cover the surface with sealing wax. Take care that the wax is uniformly melted and is in a co-herent piece, otherwise portions are apt to be detached in

FIG. 81.

handling the tube between the opera-tions of weighing ; the experiment may thus be nullified or rendered inexact. The ends should then be closed by short pieces of caoutchouc tube stopped with glass rod.

3. *The Potash Bulbs.* This appa-ratus serves to absorb the carbon di-oxide. The form represented in fig. 81 is that originally devised by Liebig (by whom, indeed, the method of organic analysis by combustion with cupric oxide

was first worked out). It is filled to the extent indicated by the dotted line in the figure, with strong potash solution, prepared by dissolving 3 parts of potash free from carbonate in 2 parts of water. The bulbs are readily filled with this liquid, contained in a porcelain dish, by dipping the end of the tube connected with the larger bulb beneath the surface of the potash solution and gently aspirating at the other tube until the required amount has been introduced. Carefully dry the tube, inside and out, with paper, and close the apparatus by short caoutchouc tubes fitted with glass rod. ·Twist a piece of platinum wire round the tubes where they touch, in the manner seen in the figure: this serves to suspend the bulbs from the hook of the balance-pan.

The tube connected with the smaller bulb is adapted to a short and light drying tube, *c* (fig. 78); about 5 centimetres long, filled with soda-lime contained between loose plugs of cotton-wool, as in the calcium chloride tube. The cork is to be trimmed and covered with sealing wax in the manner already described. This apparatus serves to retain any carbon dioxide which may escape absorption in the potash bulbs: it is therefore weighed with the bulbs.

FIG. 82.　　　FIG. 83. ,

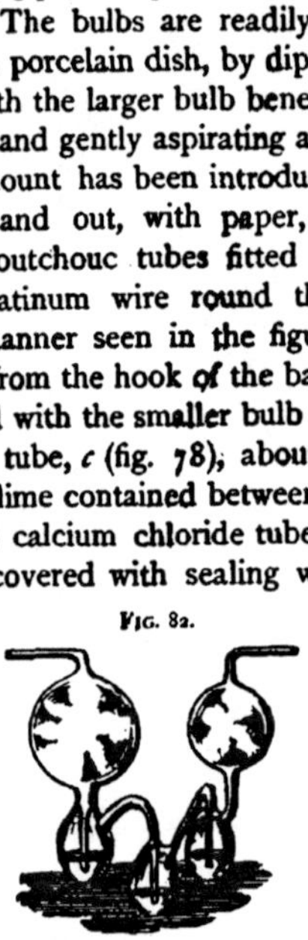

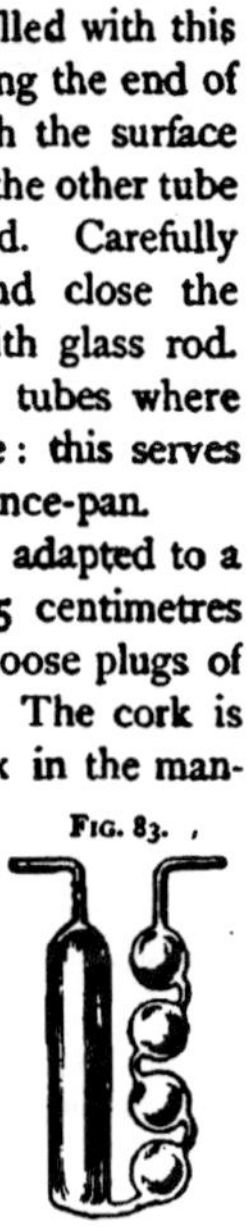

Wipe the potash bulbs and the calcium chloride tube with a soft clean cloth and place them in the balance-case.

Many other forms of the potash apparatus have been described. Fig. 82 represents a modification due to Geissler: it will be seen that the gas passes thrice through the potash solution. The apparatus requires no support and is readily filled and emptied. Fig. 83 shows a very simple form of potash bulbs, originally devised by Mitscherlich, and modified by De Koninck. This piece of apparatus is

admirably adapted for washing or drying gases.   Carbon
dioxide may also be absorbed by soda-lime, as we have frequently had occasion to observe. This method of absorption
is especially convenient if the carbon dioxide is mixed with
comparatively large quantities of other gases.   In such a
case the potash apparatus is replaced by a U-tube filled with
soda-lime and calcium chloride, as described on p. 87.

4. *A Platinum Boat,* to contain the substance to be
analysed. This should be of such size as to pass readily
into the tube. It may conveniently be 70 mm. long and
8 mm. deep.

5. *Cupric Oxide.* Strongly heat some clean copper scales
in a muffle, and when they are sufficiently cool, transfer them
to a porcelain basin and heat them with nitric acid (sp. gr.
1·2). Evaporate the pasty mass to dryness on a sand-bath,
pound it up and heat it strongly in a covered Hessian
crucible.   Break the crucible, carefully remove any pieces of
clay, and coarsely powder the fused cupric oxide, pass the
powder through a sieve of wire gauze, to separate the fine
portions. The cupric oxide to be used in the analysis should
be in little pieces about the size of hemp-seed. The finer portion should be preserved in a stoppered bottle : it is useful
for the determination of nitrogen, as described hereafter.

6. *Copper Gauze and Wire.*   Roll two pieces of fine wire
gauze, about 2 centimetres broad, into plugs of a size just
sufficient to pass easily, but with a little friction, into the
combustion-tube.   Heat them in the Bunsen flame to remove any adhering greasy matter, and when cold push one
of them down about 25 centimetres into the combustion-
tube, and fill up the tube from the other end with the
coarsely-powdered cupric oxide, occasionally tapping it so
as to shake the pieces as closely together as possible. When
the tube is filled to within 6 centimetres of the end, insert
the second plug of metallic copper. The layer of copper
oxide should be about 54 centimetres in length : there is no
necessity to leave a channel above it for the gases, since from

the coarseness of the powder there is ample room for their escape. Over the end of the tube is placed a small circular disc of copper (fig. 78), readily moveable along the tube. This serves to protect the caoutchouc stopper and to shield the little bulb *a* of the calcium chloride tube from the heat : by moving it backwards or forwards along the tube, as occasion requires, the condensation within the tube of the water produced in the combustion may be entirely prevented.

Cut another piece of the copper gauze, about 10 centimetres broad, and of the same length as you have found suitable for the plugs, and roll it round a piece of stout , copper wire, about 12 centimetres long ; the one end of the wire should be bent sharply upon itself so as to hold the copper gauze firmly near one corner : the gauze is then turned over the wire along its entire length and wrapped round so as to form a cylinder, which easily passes into the combustion-tube. The other end of the wire should be bent so as to form a little ring of less diameter than the tube. In the combustion this long cylinder of gauze is placed behind the platinum boat containing the substance to be analysed. The vacant space in the tube, that is, the portion before the first copper plug, should be sufficiently large to hold the platinum boat and copper cylinder, and still leave room for the insertion of the cork.

7. *An Apparatus for drying and removing Carbon Dioxide from the Air and from Oxygen.*—This may be arranged as in fig. 78 : the lower neck of the cylinder is partially closed by a few fragments of glass, and the cylinder is half filled with soda-lime in coarse fragments : over this is placed a layer of cotton-wool, and the remainder is filled with calcium chloride in loose spongy pieces. The cylinder is closed by a caoutchouc stopper carrying a bent tube and leading to the two large U-tubes (*a*) and (*a*), also filled with calcium chloride. Before entering the cylinder, the air or oxygen traverses the wash-bottle *b*, containing strong solution of caustic potash. This removes the greater portion of any accompanying carbon

dioxide, and also serves to indicate the speed with which the gas passes into the combustion-tube.

8. *A Bell-jar fitted with a Cork and Calcium Chloride Tube.* The bell-jar stands in a vessel containing water. By connecting it with the potash apparatus in the manner seen in fig. 78 the pressure within the combustion-tube is decreased in proportion to the height of the column of water within the bell-jar above the level of that in the trough.

The Process.—When everything is arranged, *gently* heat the combustion-tube, having previously removed the platinum boat, and pass a slow current of dry air over the copper oxide to expel any hygroscopic moisture. Whilst the oxide is being heated, weigh the potash bulbs and calcium chloride tube *without their caoutchouc stoppers*: when you have determined their weight, replace the stoppers. In about 10 or 15 minutes turn down the flames beneath the tube and allow the oxide to cool in the current of dry air. We will assume, by way of example, that you are about to analyse pure cane-sugar. This should be previously powdered and dried in the steam-bath. Heat the platinum boat to redness and allow it to cool in the desiccator. Weigh it and transfer about 0·4 grm. of the sugar into the boat, and again weigh : the increase in the weight of the boat shows the amount taken for analysis. The sugar may be accurately weighed in this manner, as it is not hygroscopic. Stop the air-current, and adapt the weighed calcium chloride tube to the combustion-tube by means of the caoutchouc stopper, and connect the potash bulbs, by a piece of well-fitting caoutchouc tubing, with the calcium chloride tube in the manner seen in fig. 78. There is no necessity to bind the caoutchouc to the glass tubes, since the reduced pressure within the apparatus, caused by the column of water in the bell-jar, effectually prevents leakage. Partially fill the bell-jar with water by aspirating at the end of the tube, turn the stopcock so as to prevent the entrance of the air, and connect the caoutchouc tubing with the end of the soda-lime

tube of the potash apparatus (fig. 78). Remove the stopper
at the further end of the combustion-tube, withdraw the
cylinder of copper gauze (which will now be superficially
oxidised), insert the platinum boat containing the weighed
amount of the sugar, and replace the cylinder, pushing the
boat nearly to the plug of copper gauze. Again fit in the
cork and connect the caoutchouc tube of the wash-bottle (*b*)
with the gasometer containing the oxygen. Now *cautiously*
open the stopcock of the bell-jar and incline the potash
bulbs in the manner seen in the figure: the smaller bulb
should be about half filled with the potash solution. Light
the first 6 or 8 burners (beginning at the end nearest the
calcium chloride tube), and gradually heat the tube. As it
becomes red-hot, light successive burners until it is at a dull
red heat. Now light the last two or three burners at the
other end of the tube, immediately under the gauze cylinder,
so as to heat it gently, and turn on a slow stream of oxygen
(about a bubble every two seconds suffices at the commencement of the process). Continue to ignite successive burners
so as to heat fresh portions of the copper oxide: when the
tube is at a dull red heat to within 4 or 5 cm. of the
platinum boat, turn on the gas in one of the burners immediately underneath the boat, and gently heat it. The sugar
will quickly melt, become brown, and give off vapours.
Carefully observe the movements of the liquid in the potash
bulbs, and regulate the heat so as to preserve a uniform
passage of gas into the bulbs.

As soon as the sugar in the boat appears to be completely
charred, and the amount of carbon dioxide passing into the
bulbs becomes small, increase the heat beneath the boat (by
this time the whole of the burners should be lighted), and
send a slightly brisker current of the dry oxygen (about one
bubble per second) through the apparatus. The carbonaceous matter within the boat gradually burns: as soon as
it has disappeared gradually diminish the flames underneath
the gauze cylinder and platinum boat, turn on a little more

oxygen, and when the gas appears to pass unabsorbed through the potash bulbs gradually lower the flames along the entire length of the tube. Close the caoutchouc tube of the wash-bottle, and transfer it to the gasometer of air, and send a current of air through the tube to displace the oxygen. In a few minutes disconnect the potash bulbs and calcium chloride tube (taking care to hold the latter so that the water condensed in the smaller bulb does not flow out), fit in their respective stoppers, wipe them, re-weigh them (of course without the stoppers). Allow the combustion-tube to cool gradually: if care be taken to anneal it properly it will serve a great number of times without rearrangement. The heat need not be so high as to distort the tube : the great majority of carbonaceous substances, especially if they contain oxygen, burn with comparative ease in contact with copper oxide and free oxygen. The apparatus is ready for a second combustion : if the analysis of the sugar has to be repeated it is of course not necessary to wait until the copper oxide and tube are completely cold.

In the analysis of volatile liquids the substance is weighed out in little bulbs of the shape seen in fig. 84. These are

FIG. 84.

made from tube, obtained by drawing out a piece of wide glass tubing before the blowpipe until it is about 5 milli-

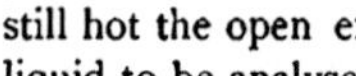

metres in external diameter. The tube to contain the liquid should be about 30 millimetres in length in the wider portion : the narrow portion should be short enough to allow the tube to rest in the platinum boat. The tube is weighed and passed once or twice through the Bunsen flame, and whilst still hot the open end is plunged beneath the surface of the liquid to be analysed. As the tube cools the liquid is driven into it to replace the air expelled on warming. Withdraw the tube from the liquid and cause the small portion within the bulb to boil briskly so as to drive out all the air, and again plunge the end of the tube into the liquid. The bulb will now be almost completely filled. Except in the case of very volatile liquids it is unnecessary to seal the end of the capillary tube : the bulb containing the liquid may be accurately weighed and transferred to the combustion-tube without any appreciable loss from evaporation. The process of combustion does not differ in any essential particulars from that already described. It is advisable, however, to expel the liquid from the bulb before the copper oxide is heated in its vicinity. As soon as the copper oxide is red hot to within 15 centimetres of the end of the platinum boat, remove one of the heated clay plates, and place it immediately over the boat : if the liquid is moderately volatile it will be readily and gradually expelled. If not, the bulb must be heated by a very small flame. The combustion of volatile liquids demands great care and attention ; the operation must not be hurried, or portions will escape unburnt.

II. ANALYSIS OF ORGANIC SUBSTANCES CONTAINING NITROGEN.

The determination of the several elements contained in an azotised organic compound cannot be very conveniently made in a single operation. It is usually preferred to estimate the carbon and hydrogen in one portion, and to determine the nitrogen in a second quantity. Nitrogenous organic substances when burnt with copper oxide, particularly if free oxygen be present, are apt to evolve nitroxygen compounds, which condense in the calcium chloride tube and potash bulbs, and vitiate the results of the carbon and hydrogen determinations. By passing the mixed products of combustion over heated metallic copper the nitroxygen compounds are decomposed; the oxygen combines with the copper, and the nitrogen passes unabsorbed through the apparatus. In the combustion of organic substances containing nitrogen it is necessary, therefore, to introduce a cylinder of copper gauze, about 12 centimetres long, rolled on a stout copper wire, exactly like that placed in the posterior part of the tube behind the platinum boat. This is to be kept at a bright red heat during the operation : the carbon and hydrogen may then be determined accurately in the ordinary way ; or a length of from four to six inches of a mixture of potassium dichromate and manganic oxide is placed in the fore part of the tube and kept at a temperature of about 250°. This mixture readily absorbs nitroxygen fumes.

Determination of Nitrogen by Volume. Maxwell Simpson's Method.—This process is applicable to all nitrogenous bodies, inorganic and organic. The substance is burnt by a mixture of cupric and mercuric oxides in a tube from which the air has previously been expelled by a current of carbon dioxide : the nitrogen and carbon dioxide, together with the excess of free oxygen, are passed over strongly-heated metallic copper, which retains the latter gas : the remaining gases are collected in an apparatus standing over mercury and partially filled with strong solution of caustic potash, which absorbs the carbon dioxide : the residual nitrogen is transferred to a measuring tube standing over mercury, and its volume is accurately determined. From the known weight of a litre of nitrogen the weight of the gas is readily calculated.

A piece of strong combustion-tube about 80 centimetres long is sealed and rounded at one end like a test-tube. A mixture of 12 grams of manganous carbonate or magnesite dried at 100°, and 2 grams of precipitated mercuric oxide are introduced into the tube. Insert a plug of recently-ignited asbestos, pushing it down to within 2 centimetres from the mixture, and afterwards add about 1 gram of the mercuric oxide. Weigh out about 0·6 gram of the nitrogenous substance to be analysed into a glazed porcelain mortar, and mix it with about 45 times its weight of a previously-prepared mixture of 4 parts of finely-powdered and recently-ignited cupric oxide and 5 parts of the dried mercuric oxide. Transfer the mixture to the tube without loss, and rinse the mortar with a fresh portion of the two oxides, adding the rinsings to the tube. Push down a second and thick plug of asbestos to within about 30 centimetres from the first, and then a layer, about 9 centimetres in length, of pure cupric oxide; next a third asbestos plug, and lastly a layer not less than 20 centimetres long of metallic copper, prepared by reducing granular cupric oxide in a stream of carbon monoxide. Draw out the end of the combustion-tube before the blowpipe and connect it with the bent delivery tube *a*, which dips beneath the surface of the mercury in the trough (fig. 85). Before placing the tube in the furnace tap it gently on the table to shake down the several layers, in order to leave a channel for the escaping gases.

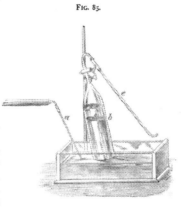

FIG. 85.

The vessel *b* has a capacity of about 200 c.c. It is provided with a glass stopcock and bent delivery tube as represented in fig. 85. To ascertain if the stopcock is perfectly air-tight, fill the apparatus completely with mercury and place it on its foot: any leakage will immediately reveal itself by the mercury flowing out of the tubulus. If the stopcock is found to be tight replace about 20 c.c. of the mercury with a strong solution of caustic potash, and place the vessel in the mercurial trough with its tubulus beneath the surface of the metal.

Heat a portion, say the posterior half of the manganous carbonate or magnesite, and drive out the air within the tube by a brisk current of carbon dioxide. At the same time commence to heat the portion of the tube occupied by the metallic copper and pure cupric oxide. As soon as the escaping gas is free from air (which is readily ascertained by allowing a quantity to pass into a test-tube filled with potash solution, when no bubble should be left), and the anterior portion of the tube is well heated, insert the end of the delivery tube through the tubulus of the vessel, and gradually heat successive portions of the tube occupied by the mixture of nitrogenous substance, cupric and mercuric oxides, beginning with the part nearest to the pure cupric oxide. As soon as no further evolution of gas is observed, and the whole length of the tube (with the exception of the part occupied by the undecomposed magnesite or manganous carbonate) is at a bright red heat, heat the remainder, so as to cause a rapid evolution of carbon dioxide, by which the

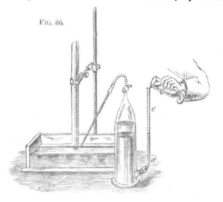

Fig. 86.

nitrogen still existing in the tube is expelled. Withdraw the delivery tube from the tubulus, and allow the gas in *b* to remain over the caustic potash solution for about an hour to absorb the last traces of the carbon dioxide. The pure nitrogen has now to be transferred to a measuring tube in order that its volume may be determined. The bent tube *c*, fig. 86, which is contracted at *d*, is fitted by the aid of a caoutchouc stopper into the tubulus, underneath the surface of the mercury in the trough. To prevent the possibility of any air adhering to the stopper and so finding its way into the nitrogen, it is advisable to moisten the stopper with a solution of corrosive sublimate before inserting it into the tubulus. Fill up the bent tube with mercury and remove *b* from the trough. Place a drop of water in the measuring tube, fill it with mercury, and invert it beneath the surface of the metal in the trough. Place the end of the delivery tube *e* beneath the measuring tube, cautiously turn the stopcock, and allow the gas to escape from *b*. When the level of the mercury in *c* approaches the contracted portion, close the

stopcock, refill the tube with the metal, and reopen the stopcock, and so gradually transfer the nitrogen into the measuring tube, closing the stopcock as soon as the potash solution touches it. Of course the delivery tube is thus left filled with nitrogen, but as an identical volume of air (viz. that which originally filled it) has been transferred to the measuring tube, no error is committed. Read off the volume of the moist gas and correct it for pressure, tension of aqueous vapour, and temperature, and calculate the weight of the nitrogen : a litre of nitrogen under the standard conditions of temperature and pressure weighs 1·255 gram.

FIG. 87.

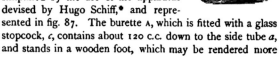

The above method of measuring the volume of the nitrogen may be much simplified by the use of the apparatus devised by Hugo Schiff,* and represented in fig. 87. The burette A, which is fitted with a glass stopcock, *c*, contains about 120 c.c. down to the side tube *a*, and stands in a wooden foot, which may be rendered more

---

* Fresenius, Zeitschrift für anal. Chemie, p. 430. 1868

stable by being weighted with lead. At about 2 centimetres beneath the side tube *a*, is a second tubulus, *b*, inclined upwards in the manner seen in the figure. Through this tube is poured mercury to a height of 2 or 3 millimetres above the lower opening. The vessel B, holding from 150 to 170 c.c., is supported by a metallic ring attached to the clamp *e*: and may thus be readily placed at any desired height along the burette: B is connected by a strong caoutchouc tubing, previously soaked in melted paraffin, with the side tube *a*. B is filled with a strong solution of potassium hydrate of sp. gr. 1·5, prepared by dissolving potash in an equal weight of water: its neck is closed by a cork, in which a narrow opening is cut. On closing the tubulus *b* with a cork, and on opening the stopcock and raising B, the potash solution flows over into the burette and completely fills it. The stopcock is now closed and the vessel B is lowered nearly to the foot of the burette : the stopper may then be withdrawn from *b* without the mercury being forced out.

The delivery tube of the combustion-tube is then pushed through *b* as soon as all the air has been expelled. The volume of the nitrogen is then directly measured, the vessel B being raised until the levels of the potash solution in both pieces of the apparatus are coincident. The nitrogen may without sensible error be assumed to be dry: the amount of moisture present in it is probably never more than 0·007 of its volume. This additive quantity serves in some measure to compensate for the deficit in the amount of nitrogen obtained, due to the impossibility of entirely preventing the formation of nitroxygen compounds in the process of combustion.

Estimation of Nitrogen as Ammonia.—By Burning with Soda-lime.—This process, which is applicable to all nitrogenous substances excepting the so-called nitro-compounds, *e.g.* nitro-benzol, amyl nitrate, &c., has already been described on p. 334.

III. ANALYSIS OF ORGANIC SUBSTANCES CONTAINING
CHLORINE, BROMINE, AND IODINE.

When an organic compound containing a halogen, chlorine,
for example, is burnt with cupric oxide, cuprous chloride is
formed, which, being volatile, is carried forward in the
stream of gas and condenses in the calcium chloride tube,
and thus renders the determination of the hydrogen inexact. If the gases within the tube contain free oxygen
the cuprous chloride is more or less decomposed, cupric
oxide being formed, and chlorine eliminated. This is
retained partly by the calcium chloride, partly by the
potash solution. By inserting a cylinder of copper gauze
in the anterior portion of the tube, the chlorine may be
arrested so long as the amount of oxygen is not sufficient to
oxidise the copper. By mixing the cupric oxide with a
small quantity of lead oxide the chlorine may be entirely
retained.

The determination of the carbon and hydrogen in compounds containing chlorine is best effected by heating with
lead chromate. This substance is readily made by mixing
potassium chromate and lead nitrate or acetate solutions,
thoroughly washing the dense yellow precipitate, drying it,
heating it to redness in a covered clay crucible, and coarsely
powdering it. The combustion is made in the manner
already described in the case of copper oxide.

Determination of the Halogen.—A narrow piece of combustion-tube about 40 centimetres long is sealed and rounded
at the end like a test-tube. A small quantity of coarsely-
powdered and recently-burnt lime is introduced into it, so as
to occupy a length of 4 centimetres. The compound to be
analysed, if solid, is weighed out into the combustion-tube,
and mixed with a quan-
tity of the lime in mo-
derately fine powder,
by the aid of a brass wire bent in the manner seen in fig. 88.

FIG. 88.

By twisting this wire among the fragments the substance and the lime are uniformly mixed. The wire is rinsed from any adhering powder by a further quantity of lime, and the tube is filled with the coarsely-powdered lime to within about 3 or 4 centimetres from the open end, placed in the furnace and closed by a cork carrying a short piece of bent tube, which dips beneath the surface of water contained in a small beaker. This serves to maintain a slight pressure within the tube and tends to prevent the escape of any of the halogen. Commence the operation by heating the anterior portion of the tube, and gradually approach the part containing the substance as the lime becomes red-hot. Having lighted all the burners beneath it, continue to heat the tube until the cessation of gas bubbling through the water tells you that the process is finished. When the tube is cold, empty the loose fragments of the lime into about 150 c.c. of water, and half fill the tube with water to dissolve any fused substance adhering to the glass. Acidify the liquid with moderately dilute nitric acid free from chlorine : an excess of nitric acid is readily indicated by the change in the colour of the suspended carbonaceous matter. Immediately all the lime is dissolved the precipitate becomes quite black. The liquid is filtered and treated with silver nitrate solution, and the precipitated silver salt washed, dried, and weighed. Of course the quantity of the chlorine may be estimated volumetrically by standard silver solution and potassium chromate if care be taken to neutralise the excess of nitric acid by well-washed precipitated calcium carbonate, or by the addition of sodium carbonate solution.

Liquids containing chlorine, &c., are weighed out in bulbs, as described on p. 354 : after the introduction of a layer of lime, about 4 centimetres long, the bulb is allowed to slide down the tube, which is then immediately filled up with lime. When about half the length of the tube has been heated, expel the liquid from the bulb by gently heating the tube where it is situated, and conduct the remainder of the operation as described.

Many organic substances containing a halogen may be very conveniently analysed by digesting them with water and sodium amalgam. The liquid poured off the residual mercury is acidified with nitric acid and the chlorine determined in the usual manner.

Certain organic iodides are decomposed by heating them with an alcoholic solution of silver nitrate: the silver iodide thus formed may be filtered off, dried, and weighed.

IV. ANALYSIS OF ORGANIC SUBSTANCES CONTAINING SULPHUR AND PHOSPHORUS.

The combustion of organic bodies containing sulphur is most accurately made with lead chromate: the only pre-caution needed is to maintain the anterior portion of the tube, to the extent of 15 or 20 centimetres, at a very low red heat only. Under these circumstances no sulphur dioxide passes into the absorption apparatus.

Determination of Sulphur.—Solid substances containing sulphur may be decomposed by fusion with potassium hydrate and pure nitre. Place a quantity of potassium hydrate in a silver dish, mix it with about ⅓ of its weight of nitre and fuse the mixture. Allow it to cool and add to it the weighed quantity of the sulphur compound. Heat gently, and stir continually with a silver spatula, adding little by little a small quantity of nitre if the carbon appears to be but slowly con-sumed. When the mass is cold, dissolve it in water, acidify with hydrochloric acid, boil, and add barium chloride. Treat the precipitated barium sulphate in the manner de-scribed on p. 169.

Solid compounds of sulphur may also be analysed by digesting them with strong potash solution contained in a large porcelain crucible, and passing a stream of chlorine into the liquid until the substance is completely decomposed. Acidify, heat gently to expel excess of chlorine, filter, and add barium chloride.

Carius' Method. Applicable to the Estimation of Sulphur and Phosphorus in Solid and Liquid Substances.—The compound is oxidised by the action of nitric acid of sp. gr. 1·2. From o·2 gram to o·4 gram of the substance is weighed out in a thin glass-bulb, care being taken that but little air is enclosed within the bulb. The sealed bulb is brought into a tube of hard glass of about 10 or 12 millimetres in internal diameter, sealed and rounded at one end like a test-tube, together with from 20 to 60 times its weight of nitric acid of sp. gr. 1·2. The tube must not be more than half filled with the liquid. It is now softened in the blowpipe flame at a few centimetres from the open end, and the fused glass allowed to thicken, and it is then drawn out into a thick-walled capillary tube. The tube is supported in the clamp of a retort stand, and the nitric acid caused to boil, so as to expel the air contained within the tube : when the acid vapours are freely evolved, the lamp is removed, and the capillary opening is closed by the blowpipe flame. Allow the liquid to become nearly cold, wrap the tube in a thick towel (for safety), and break the bulb by shaking it smartly against the ends of the tube. Heat the tube to 120-150° for some hours in the air-bath. Allow the bath to cool before withdrawing the tube, wrap it in the towel, and cautiously warm the point, so as to expel the liquid which collects in the capillary tube. Soften the end in the blowpipe flame : the enclosed gases will force their way through the fused glass. Examine the tube carefully, and if you have reason to believe that the oxidation is incomplete, re-seal the tube and heat it to 180° for an hour. Allow it to cool and open it with the same precautions as before. If no more gas escapes the process is finished. Cut off the end of the tube, rinse its contents into a beaker, dilute with water, and, in the case of sulphur, add barium chloride. In the case of phosphorus, add ammonia, ammonium chloride, and magnesia mixture, and convert the precipitate into magnesium pyrophosphate. If sulphur and phosphorus are together present, precipitate the sulphuric acid with barium chloride, remove the excess of baryta by sulphuric acid, concentrate the filtrate by evaporation, and determine the phosphoric acid as magnesium pyrophosphate.

APPENDIX.

—•◦•—

TABLE I.

Symbols and Atomic Weights of the Elements.

Element	Symbol	Atomic weight	Observer
Aluminium .	Al	27·02	Mallet
Antimony .	Sb	119·6	Schneider ; Cooke
Arsenic . .	As	74·9	Kessler
Barium . .	Ba	136·84	Marignac
Bismuth . .	Bi	207·5	Dumas
Boron . .	B	10·9	Berzelius
Bromine .	Br	79·76	Stas
Cadmium .	Cd	111·7	Lenssen
Cæsium . .	Cs	132·7	Johnson and Allen ; Bunsen
Calcium .	Ca	39·90	Erdmann and Marchand
Carbon .	C	11·97	Dumas and Stas ; Liebig
Cerium .	Ce	138·24	Rammelsberg
Chlorine .	Cl	35·37	Stas
Chromium .	Cr	52·08	Siewert
Cobalt .	Co	58·6	Russell
Copper .	Cu	63·12	Millon and Commaille
Didymium .	D	142·44	Hermann
Erbium .	E	168·9	Bahr and Bunsen
Fluorine .	F	18·96	Luca ; Louyet
Gallium .	Ga	69·8	Lecoq de Boisbaudran
Glucinum .	Gl	9·30	Awdejew ; Klatzo
Gold .	Au	196·85	Thorpe and Laurie
Hydrogen .	H	1	Dulong and Berzelius
Indium .	In	113·4	Winkler ; Bunsen
Iodine .	I	126·54	Stas
Iridium .	Ir	192·5	Seubert

TABLE I.—*continued.*

Element	Symbol	Atomic weight	Observer
Iron . .	Fe	55·9	Dumas
Lanthanum .	La	139·33	Hermann
Lead . .	Pb	206·40	Stas
Lithium . .	Li	7·00	Stas
Magnesium .	Mg	23·94	Dumas
Manganese .	Mn	54·8	Dewar and Scott
Mercury . .	Hg	199·8	Erdmann and Marchand
Molybdenum .	Mo	95·9	Dumas ; Debray
Nickel . .	Ni	58·6	Russell
Niobium . .	Nb	93·7	Marignac
Nitrogen . .	N	14·01	Stas
Osmium . .	Os	190·8	Seubert
Oxygen . .	O	15·96	Nilson
Palladium .	Pd	106·2	Berzelius
Phosphorus .	P	30·96	Schrötter
Platinum . .	Pt	194·38	Seubert
Potassium .	K	39·04	Stas
Rhodium .	Rh	104·1	Berzelius
Rubidium .	Rb	85·2	Bunsen ; Piccard
Ruthenium .	Ru	103·5	Berzelius
Scandium .	Sc	44·0	Nilson
Selenium . .	Se	78·9	Dumas
Silver . .	Ag	107·67	Stas
Silicon . .	Si	28·33	Thorpe and Young
Sodium . .	Na	22·99	Stas
Strontium .	Sr	87·34	Marignac
Sulphur . .	S	31·996	Stas
Tantalum .	Ta	182·00	Marignac
Tellurium .	Te	125·0	Brauner
Thallium .	Tl	203·50	Crookes
Thorium . .	Th	231·44	Delafontaine
Tin . . .	Sn	117·4	Dumas
Titanium .	Ti	48·0	Thorpe
Tungsten .	W	183·6	Schneider ; Dumas; Roscoe
Uranium . .	U	239·8	Ebelmen
Vanadium .	V	51·0	Roscoe
Ytterbium .	Yt	173·0	Nilson
Yttrium . .	Y	88·9	Bahr and Bunsen
Zinc . .	Zn	64·7	Axel Erdmann
Zirconium .	Zr	90·4	Marignac ; Bailey

TABLE II.

Volume and Density of Water at different Temperatures.

(Mean results of the observations of Kopp, Pierre, Despretz, Hagen,
Matthiessen, Weidner, Kremers, and Rossetti.)

Temp.	Sp. gr. of Water (at 0° = 1)	Vol. of Water (at 0° = 1)	Sp. gr. of Water (at 4° = 1)	Volume of Water (at 4° = 1)
0	1·000000	1·000000	·999871	1·000129
1	1·000057	0·999943	·999928	1·000072
2	1·000098	·999902	·999969	1·000031
3	1·000120	·999880	·999991	1·000009
4	1·000129	·999871	1·000000	1·000000
5	1·000119	·999881	0·999990	1·000010
6	1·000099	·999901	·999970	1·000030
7	1·000062	·999938	·999933	1·000067
8	1·000015	·999985	·999886	1·000114
9	0·999953	1·000047	·999824	1·000176
10	·999876	1·000124	·999747	1·000253
11	·999784	1·000216	·999655	1·000345
12	·999678	1·000322	·999549	1·000451
13	·999559	1·000441	·999430	1·000570
14	·999429	1·000572	·999299	1·000701
15	·999289	1·000712	·999160	1·000841
16	·999131	1·000870	·999002	1·000999
17	·998970	1·001031	·998841	1·001160
18	·998782	1·001219	·998654	1·001348
19	·998588	1·001413	·998460	1·001542
20	·998388	1·001615	·998259	1·001744
21	·998176	1·001828	·998047	1·001957
22	·997953	1·002049	·997826	1·002177
23	·997730	1·002276	·997601	1·002405
24	·997495	1·002511	·997367	1·002641
25	·997249	1·002759	·997120	1·002888
26	·996994	1·003014	·996866	1·003144
27	·996732	1·003278	·996603	1·003408
28	·996460	1·003553	·996331	1·003682
29	·996179	1·003835	·996051	1·003965
30	·995894	1·004123	·995765	1·004253
35	·99431	1·00572	·99418	1·00586
40	·99248	1·00757	·99235	1·00770
50	·98833	1·01181	·98820	1·01195
60	·98351	1·01677	·98338	1·01691
70	·97807	1·02243	·97794	1·02256
80	·97206	1·02874	·97194	1·02887
90	·96568	1·03554	·96556	1·03567
100	·95878	1·04300	·95865	1·04312

TABLE

Tension of Aqueous Vapour in

°	Mm.	°	Mm.	°	Mm.	°	Mm.	°	Mm.	°	Mm.	°	Mm.
0·0	4·600	2·5	5·491	5·0	6·534	7·5	7·751	10·0	9·165	12·5	10·804	15·0	12·699
·1	·633	·6	·530	·1	·580	·6	·804	·1	·227	·6	·875	·1	·781
·2	·667	·7	·569	·2	·625	·7	·857	·2	·288	·7	·947	·2	·864
·3	·700	·8	·608	·3	·671	·8	·910	·3	·350	·8	11·019	·3	·947
·4	·733	·9	·647	·4	·717	·9	·964	·4	·412	·9	·090	·4	13·029
·5	·767	3·0	·687	·5	·763	8·0	8·017	·5	·474	13·0	·162	·5	·112
·6	·801	·1	·727	·6	·810	·1	·072	·6	·537	·1	·235	·6	·197
·7	·836	·2	·767	·7	·857	·2	·126	·7	·601	·2	·309	·7	·281
·8	·871	·3	·807	·8	·904	·3	·181	·8	·665	·3	·383	·8	·366
·9	·905	·4	·848	·9	·951	·4	·236	·9	·728	·4	·456	·9	·451
1·0	·940	·5	·889	6·0	·998	·5	·291	11·0	·792	·5	·530	16·0	·536
·1	·975	·6	·930	·1	7·047	·6	·347	·1	·857	·6	·605	·1	·623
·2	5·011	·7	·972	·2	·095	·7	·404	·2	·923	·7	·681	·2	·710
·3	·047	·8	6·014	·3	·144	·8	·461	·3	·989	·8	·757	·3	·797
·4	·082	·9	·055	·4	·193	·9	·517	·4	10·054	·9	·832	·4	·885
·5	·118	4·0	·097	·5	·242	9·0	·574	·5	·120	14·0	·908	·5	·972
·6	·155	·1	·140	·6	·292	·1	·632	·6	·187	·1	·986	·6	14·062
·7	·191	·2	·183	·7	·342	·2	·690	·7	·255	·2	12·064	·7	·151
·8	·228	·3	·226	·8	·392	·3	·748	·8	·322	·3	·142	·8	·241
·9	·265	·4	·270	·9	·442	·4	·807	·9	·389	·4	·220	·9	·331
2·0	·302	·5	·313	7·0	·492	·5	·865	12·0	·457	·5	·298	17·0	·421
·1	·340	·6	·357	·1	·544	·6	·925	·1	·526	·6	·378	·1	·513
·2	·378	·7	·407	·2	·595	·7	·985	·2	·596	·7	·458	·2	·605
·3	·416	·8	·445	·3	·647	·8	9·045	·3	·665	·8	·538	·3	·697
·4	·454	·9	·490	·4	·699	·9	·105	·4	·734	·9	·619	·4	·790

III.

Millimetres of Mercury from 0° to 34·9° C.

°	Mm.	°	Mm.	°	Mm.	°	Mm.	°	Mm.	°	Mm.	°	Mm.
17·5	14·882	20·0	17·391	22·5	20·265	25·0	23·550	27·5	27·294	30·0	31·548	32·5	36·370
·6	·977	·1	·500	·6	·389	·1	·692	·6	·455	·1	·729	·6	·576
·7	15·072	·2	·608	·7	·514	·2	·834	·7	·617	·2	·911	·7	·783
·8	·167	·3	·717	·8	·639	·3	·976	·8	·778	·3	32·094	·8	·991
·9	·262	·4	·826	·9	·763	·4	24·119	·9	·939	·4	·278	·9	37·200
18·0	·357	·5	·935	23·0	·888	·5	·261	28·0	28·101	·5	·463	33·0	·410
·1	·454	·6	18·047	·1	21·016	·6	·406	·1	·267	·6	·650	·1	·621
·2	·552	·7	·159	·2	·144	·7	·552	·2	·433	·7	·837	·2	·832
·3	·650	·8	·271	·3	·272	·8	·697	·3	·599	·8	33·026	·3	38·045
·4	·747	·9	·383	·4	·400	·9	·842	·4	·765	·9	·215	·4	·258
·5	·845	21·0	·495	·5	·528	26·0	·988	·5	·931	31·0	·405	·5	·473
·6	·945	·1	·610	·6	·659	·1	25·138	·6	29·101	·1	·596	·6	·689
·7	16·045	·2	·724	·7	·790	·2	·288	·7	·271	·2	·787	·7	·906
·8	·145	·3	·839	·8	·921	·3	·438	·8	·441	·3	·980	·8	39·124
·9	·246	·4	·954	·9	22·058	·4	·588	·9	·612	·4	34·174	·9	·344
19·0	·346	·5	19·069	24·0	·184	·5	·738	29·0	·782	·5	·368	34·0	·565
·1	·449	·6	·187	·1	·319	·6	·891	·1	·956	·6	·564	·1	·786
·2	·552	·7	·305	·2	·453	·7	26·045	·2	30·131	·7	·761	·2	40·007
·3	·655	·8	·423	·3	·588	·8	·198	·3	·305	·8	·959	·3	·230
·4	·758	·9	·541	·4	·723	·9	·351	·4	·479	·9	35·159	·4	·455
·5	·861	22·0	·659	·5	·858	27·0	·505	·5	·654	32·0	·359	·5	·680
·6	·967	·1	·780	·6	·996	·1	·663	·6	·833	·1	·559	·6	·907
·7	17·073	·2	·901	·7	23·135	·2	·820	·7	31·011	·2	·760	·7	41·135
·8	·179	·3	20·022	·8	·273	·3	·978	·8	·190	·3	·962	·8	·364
·9	·285	·4	·143	·9	·411	·4	27·136	·9	·369	·4	36·165	·9	·595

Table IV.

BAUMÉ'S HYDROMETER.

Table for Liquids heavier than Water.

Degrees Baumé	Sp. gr.	°B.	Sp. gr.	°B.	Sp. gr.
0	1·000	26	1·206	52	1·520
1	1·007	27	1·216	53	1·535
2	1·013	28	1·226	54	1·551
3	1·020	29	1·236	55	1·567
4	1·027	30	1·246	56	1·583
5	1·034	31	1·256	57	1·600
6	1·041	32	1·267	58	1·617
7	1·048	33	1·277	59	1·634
8	1·056	34	1·288	60	1·652
9	1·063	35	1·299	61	1·670
10	1·070	36	1·310	62	1·689
11	1·078	37	1·322	63	1·708
12	1·086	38	1·333	64	1·727
13	1·094	39	1·345	65	1·747
14	1·101	40	1·357	66	1·767
15	1·109	41	1·369	67	1·788
16	1·118	42	1·382	68	1·809
17	1·126	43	1·395	69	1·831
18	1·134	44	1·407	70	1·854
19	1·143	45	1·421	71	1·877
20	1·152	46	1·434	72	1·900
21	1·160	47	1·448	73	1·924
22	1·169	48	1·462	74	1·949
23	1·178	49	1·476	75	1·974
24	1·188	50	1·490	76	2·000
25	1·197	51	1·505		

TABLE IV.—*continued.*

Table for Liquids lighter than Water.

°B.	Sp. gr.	°B.	Sp. gr.	°B.	Sp. gr.
10	1·000	27	0·896	44	0·811
11	0·993	28	0·890	45	0·807
12	0·986	29	0·885	46	0·802
13	0·980	30	0·880	47	0·798
14	0·973	31	0·874	48	0·794
15	0·967	32	0·869	49	0·789
16	0·960	33	0·864	50	0·785
17	0·954	34	0·859	51	0·781
18	0·948	35	0·854	52	0·777
19	0·942	36	0·849	53	0·773
20	0·936	37	0·844	54	0·768
21	0·930	38	0·839	55	0·764
22	0·924	39	0·834	56	0·760
23	0·918	40	0·830	57	0·757
24	0·913	41	0·825	58	0·753
25	0·907	42	0·820	59	0·749
26	0·901	43	0·816	60	0·745

TWADDELL'S HYDROMETER.

To convert degrees Twaddell into specific gravity (water = 1,00
multiply the number by 5, and add 1,000 to the product.

To reduce specific gravity (water = 1,000) to Twaddell : ded
,000 and divide the remainder by 5.

TABLE V.

Showing the Percentages of real Sulphuric Acid (SO_4H_2) corresponding to various Specific Gravities of Aqueous Sulphuric Acid.

Bineau; Otto. Temp. 15°.

Specific gravity	Per cent.	Specific gravity	Per cent.	Specific gravity	Per cent.	Specific gravity	Per cent.
1·8426	100	1·675	75	1·398	50	1·182	25
1·842	99	1·663	74	1·3886	49	1·174	24
1·8406	98	1·651	73	1·379	48	1·167	23
1·840	97	1·639	72	1·370	47	1·159	22
1·8384	96	1·627	71	1·361	46	1·1516	21
1·8376	95	1·615	70	1·351	45	1·144	20
1·8356	94	1·604	69	1·342	44	1·136	19
1·834	93	1·592	68	1·333	43	1·129	18
1·831	92	1·580	67	1·324	42	1·121	17
1·827	91	1·568	66	1·315	41	1·1136	16
1·822	90	1·557	65	1·306	40	1·106	15
1·816	89	1·545	64	1·2976	39	1·098	14
1·809	88	1·534	63	1·289	38	1·091	13
1·802	87	1·523	62	1·281	37	1·083	12
1·794	86	1·512	61	1·272	36	1·0756	11
1·786	85	1·501	60	1·264	35	1·068	10
1·777	84	1·490	59	1·256	34	1·061	9
1·767	83	1·480	58	1·2476	33	1·0536	8
1·756	82	1·469	57	1·239	32	1·0464	7
1·745	81	1·4586	56	1·231	31	1·039	6
1·734	80	1·448	55	1·223	30	1·032	5
1·722	79	1·438	54	1·215	29	1·0256	4
1·710	78	1·428	53	1·2066	28	1·019	3
1·698	77	1·418	52	1·198	27	1·013	2
1·686	76	1·408	51	1·190	26	1·0064	1

TABLE VI.

Giving the Percentage Amount of Hydrochloric Acid contained in Aqueous Solutions of the Gas of various Specific Gravities.

Ure. Temp. 15°.

Specific gravity	HCl per cent.	Specific gravity	HCl per cent.	Specific gravity	HCl per cent.	Specific gravity	HCl per cent.
1·2000	40·777	1·1515	30·582	1·1000	20·388	1·0497	10·194
1·1982	40·369	1·1494	30·174	1·0980	19·980	1·0477	9·786
1·1964	39·961	1·1473	29·767	1·0960	19·572	1·0457	9·379
1·1946	39·554	1·1452	29·359	1·0939	19·165	1·0437	8·971
1·1928	39·146	1·1431	28·951	1·0919	18·757	1·0417	8·563
1·1910	38·738	1·1410	28·544	1·0899	18·349	1·0397	8·155
1·1893	38·330	1·1389	28·136	1·0879	17·941	1·0377	7·747
1·1875	37·923	1·1369	27·728	1·0859	17·534	1·0357	7·340
1·1857	37·516	1·1349	27·321	1·0838	17·126	1·0337	6·932
1·1846	37·108	1·1328	26·913	1·0818	16·718	1·0318	6·524
1·1822	36·700	1·1308	26·505	1·0798	16·310	1·0298	6·116
1·1802	36·292	1·1287	26·098	1·0778	15·902	1·0279	5·709
1·1782	35·884	1·1267	25·690	1·0758	15·494	1·0259	5·301
1·1762	35·476	1·1247	25·282	1·0738	15·087	1·0239	4·893
1·1741	35·068	1·1226	24·874	1·0718	14·679	1·0220	4·486
1·1721	34·660	1·1206	24·466	1·0697	14·271	1·0200	4·078
1·1701	34·252	1·1185	24·058	1·0677	13·863	1·0180	3·670
1·1681	33·845	1·1164	23·650	1·0657	13·456	1·0160	3·262
1·1661	33·437	1·1143	23·242	1·0637	13·049	1·0140	2·854
1·1641	33·029	1·1123	22·834	1·0617	12·641	1·0120	2·447
1·1620	32·621	1·1102	22·426	1·0597	12·233	1·0100	2·039
1·1599	32·213	1·1082	22·019	1·0577	11·825	1·0080	1·631
1·1578	31·805	1·1061	21·611	1·0557	11·418	1·0060	1·124
1·1557	31·398	1·1041	21·203	1·0537	11·010	1·0040	0·816
1·1536	30·990	1·1020	20·796	1·0517	10·602	1·0020	0·408

TABLE VII.

Showing the Percentage Amount of Nitric Acid (HNO_3) contained in Aqueous Solutions of various Specific Gravities.

(Kolb, Ann. Ch. Phys. [4] 136).

The numbers marked * are the results of direct observations; the others are obtained by interpolation.

HNO_3 per cent.	Specific gravity		Contraction	HNO_3 per cent.	Specific gravity		Contraction
	At 0°	At 15°			At 0°	At 15°	
100·00	1·559	1·530	0·0000	68·00	1·435	1·414	0·0784
99·84*	1·559*	1·530*	0·0004	67·00	1·430	1·410	0·0796
99·72*	1·558*	1·530*	0·0010	66·00	1·425	1·405	0·0806
99·52*	1·557*	1·529*	0·0014	65·07*	1·420*	1·400*	0·0818
97·89*	1·551*	1·523*	0·0065	64·00	1·415	1·395	0·0830
97·00	1·548	1·520	0·0090	63·59	1·413	1·393	0·0833
96·00	1·544	1·516	0·0120	62·00	1·404	1·386	0·0846
95·27*	1·542*	1·514*	0·0142	61·21*	1·400*	1·381*	0·0850
94·00	1·537	1·509	0·0182	60·00	1·393	1·374	0·0854
93·01*	1·533*	1·506*	0·0208	59·59*	1·391*	1·372*	0·0855
92·00	1·529	1·503	0·0242	58·88	1·387	1·368	0·0861
91·00	1·526	1·499	0·0272	58·00	1·382	1·363	0·0864
90·00	1·522	1·495	0·0301	57·00	1·376	1·358	0·0868
89·56*	1·521*	1·494*	0·0315	56·10*	1·371*	1·353*	0·0870
88·00	1·514	1·488	0·0354	55·00	1·365	1·346	0·0874
87·45*	1·513*	1·486*.	0·0369	54·00	1·359	1·341	0·0875
86·17*	1·507*	1·482	0·0404	53·81	1·358	1·339	0·0875
85·00	1·503	1·478	0·0433	53·00	1·353	1·335	0·0875
84·00	1·499	1·474	0·0459	52·33*	1·349*	1·331*	0·0875
83·00	1·495	1·470	0·0485	50·99*	1·341*	1·323*	0·0872
82·00	1·492	1·467	0·0508	49·97	1·334	1·317	0·0867
80·96*	1·488*	1·463*	0·0531	49·00	1·328	1·312	0·0862
80·00	1·484	1·460	0·0556	48·00	1·321	1·304	0·0856
79·00	1·481	1·456	0·0580	47·18*	1·315*	1·298*	0·0850
77·66	1·476	1·451	0·0610	46·64	1·312	1·295	0·0848
76·00	1·469	1·445	0·0643	45·00	1·300	1·284	0·0835
75·00	1·465	1·442	0·0666	43·53*	1·291*	1·274*	0·0820
74·01*	1·462*	1·438*	0·0688	42·00	1·280	1·264	0·0808
73·00	1·457	1·435	0·0708	41·00	1·274	1·257	0·0796
72·39*	1·455*	1·432*	0·0722	40·00	1·267	1·251	0·0786
71·24*	1·450*	1·429*	0·0740	39·00	1·260	1·244	0·0755
69·96	1·444	1·423	0·0760	37·95*	1·253*	1·237*	0·0762
69·20*	1·441	1·419*	0·0771	36·00	1·240	1·225	0·0740

TABLE VII.—*continued.*

HNO₃ per cent.	Specific gravity At 0°	At 15°	Contraction	HNO₃ per cent.	Specific gravity At 0°	At 15°	Contraction
35·00	1·234	1·218	0·0729	20·00	1·132	1·120	0·0483
33·86*	1·226*	1·211*	0·0718	17·47*	1·115*	1·105*	0·0422
32·00	1·214	1·198	0·0692	15·00	1·099	1·089	0·0336
31·00	1·207	1·192	0·0678	13·00	1·085	1·077	0·0316
30·00	1·200	1·185	0·0664	11·41*	1·075	1·067*	0·0296
29·00	1·194	1·179	0·0650	7·22*	1·050	1·045*	0·0206
28·00*	1·187*	1·172*	0·0635	4·00	1·026	1·022	0·0112
27·00	1·180	1·166	0·0616	2·00	1·013	1·010	0·0055
25·71*	1·171*	1·157*	0·0593	0·00	1·000	0·999	0·0000
23·00	1·153	1·138	0·0520				

TABLE VIII.

Showing the Percentage Amount of Caustic Potash (K₂O) in Aqueous Solutions of various Specific Gravities.

Tünnermann, N. Tr. xviii., 2, 5. Temp. 15°.

Sp. gr.	Per cent.	Sp. gr.	Per cent.
1·3300	28·290	1·1437	14·145
1·3131	27·158	1·1308	13·013
1·2966	26·027	1·1182	11·832
1·2805	24·895	1·1059	10·750
1·2648	23·764	1·0938	9·619
1·2493	22·632	1·0819	8·487
1·2342	21·500	1·0703	7·355
1·2268	20·935	1·0589	6·224
1·2122	19·803	1·0478	5·002
1·1979	18·671	1·0369	3·961
1·1839	17·540	1·0260	2·829
1·1702	16·408	1·0153	1·697
1·1568	15·277	1·0050	0·5658

Appendix.

TABLE IX.

Showing Percentage Amount of Soda (Na_2O) in Aqueous Solutions of various Specific Gravities.

Tünnermann.

Sp. gr.	Per cent.	Sp. gr.	Per cent.	Sp. gr.	Per cent.	Sp. gr.	Per cent.
1·4285	30·220	1·3198	22·363	1·2392	15·110	1·1042	7·253
1·4193	29·616	1·3143	21·894	1·2280	14·500	1·0948	6·648
1·4101	29·011	1·3125	21·758	1·2178	13·901	1·0855	6·044
1·4011	28·407	1·3053	21·154	1·2058	13·297	1·0764	5·440
1·3923	27·802	1·2982	20·550	1·1948	12·692	1·0675	4·835
1·3836	27·200	1·2912	19·945	1·1841	12·088	1·0587	4·231
1·3751	26·594	1·2843	19·341	1·1734	11·484	1·0500	3·626
1·3668	25·989	1·2775	18·730	1·1630	10·879	1·0414	3·022
1·3586	25·385	1·2708	18·132	1·1528	10·275	1·0330	2·418
1·3505	24·780	1·2642	17·528	1·1428	9·670	1·0246	1·813
1·3426	24·176	1·2578	16·923	1·1330	9·066	1·0163	1·209
1·3349	23·572	1·2515	16·379	1·1233	8·462	1·0081	0·604
1·3273	22·967	1·2453	15·714	1·1137	7·857	1·0040	0·302

TABLE X.

Showing the Percentage Amount of Ammonia in Aqueous Solutions of the Gas of various Specific Gravities.

Carius. Temp. 14°.

Sp. gravity	NH_3 per cent.	Sp. gravity	NH_3 per cent.	Sp. gravity	NH_3 per cent.
0·8844	36	0·9133	24	0·9520	12
0·8864	35	0·9162	23	0·9556	11
0·8885	34	0·9191	22	0·9593	10
0·8907	33	0·9221	21	0·9631	9
0·8929	32	0·9251	20	0·9670	8
0·8953	31	0·9283	19	0·9709	7
0·8976	30	0·9314	18	0·9749	6
0·9001	29	0·9347	17	0·9790	5
0·9026	28	0·9380	16	0·9031	4
0·9052	27	0·9414	15	0·9873	3
0·9078	26	0·9449	14	0·9915	2
0·9106	25	0·9484	13	0·9959	1

TABLE XI.

Reduction of Weighings in Air to a Vacuum (*G. F. Becker,*
Liebig's Annalen, 195, *p.* 222).

Brass weights for substances whose sp. gr. is between	Correction per gram, error less than $\frac{1}{10}$ mgrm.	Platinum weights for substances whose sp. gr. is between
27·738 and 11·064	− 0·000067	51·766 and 13·568
11·064 ,, 6·904	0·000000	13·568 ,, 7·807
6·904 ,, 5·019	+ 0·000067	7·807 ,, 5·480
5·019 ,, 3·943	0·000133	5·480 ,, 4·222
3·943 ,, 3·247	0·000200	4·222 ,, 3·433
3·247 ,, 2·759	0·000267	3·433 ,, 2·893
2·759 ,, 2·399	0·000333	2·893 ,, 2·500
2·399 ,, 2·122	0·000400	2·500 ,, 2·201
2·122 ,, 1·903	0·000467	2·201 ,, 1·965
1·903 ,, 1·724	0·000533	1·965 ,, 1·776
1·724 ,, 1·576	0·000600	1·776 ,, 1·619
1·576 ,, 1·452	0·000667	1·619 ,, 1·488
1·452 ,, 1·377	0·000733	1·488 ,, 1·377
1·377 ,, 1·254	0·000800	1·377 ,, 1·281
1·254 ,, 1·174	0·000867	1·281 ,, 1·197
1·174 ,, 1·103	0·000933	1·197 ,, 1·124
1·103 ,, 1·041	0·001000	1·124 ,, 1·059
1·041 ,, 0·985	0·001067	1·059 ,, 1·002
	0·001133	1·002 ,, 0·950
	0·001200	

PREPARATION OF PURE PLATINUM TETRA-CHLORIDE.

Scrap platinum, which may contain iridium, osmium, &c., is dissolved in aqua regia, and the solution is evaporated to dryness; the residue is dissolved in moderately concentrated hydrochloric acid, and again evaporated to dryness. The dried chloride is once more dissolved in hot water containing free hydrochloric acid, and the solution mixed with a large excess of soda-ley. It is again boiled for some time, and a few drops of alcohol are added in order to destroy any sodium hypochlorite which may be formed; the precipitate is redissolved in hydrochloric acid; the liquid is filtered, if necessary, and mixed with a hot and saturated solution of ammonium chloride so long as a precipitate forms. This process of separating platinum from its congeners, with which in the commercial variety of the metal it is almost invariably mixed, is based upon the different behaviour of sodium hydrate solution towards the higher chlorides of the associated metals. Platinic chloride is very slightly, if at all, reduced to platinous chloride on boiling with soda solution, whereas the other chlorides are all reduced, with the production of sodium chloride and hypochlorite, and these reduced chlorides are not precipitated in union with ammonium chloride.

The ammonium-platinum chloride, which is of a bright yellow colour and free from orange or red crystals, is washed by decantation, dried, and gently heated in a platinum crucible, or it may be placed in a piece of hard glass tubing and decomposed in a current of coal gas or dry hydrogen. The reduced metal should be weighed, dissolved in aqua regia, the solution evaporated to dryness with excess of hydrochloric acid to expel the last traces of nitric acid, and the residue dissolved at a gentle heat in a definite volume of dilute hydrochloric acid. The operator in this manner obtains an idea of the strength of the solution.

Pure platinum chloride may be recovered from the precipitates with the alkaline chlorides which are obtained in analytical work by boiling them with a solution of sodium carbonate and alcohol : washing the precipitated platinum with hot water by decantation and finally with hot hydrochloric acid. The spongy metal is then dissolved in aqua regia ($5HCl : 1HNO_3$), the solution filtered, and evaporated to dryness with hydrochloric acid as above.

TREATMENT OF 'SILVER RESIDUES.'

The mixed silver salts, associated with metallic silver, which accumulate in the course of analytical work, may be conveniently reduced, after washing and drying, by heating to fusion with a mixture of sodium and potassium carbonates in an earthen or unglazed porcelain crucible. The button of metallic silver is washed with boiling water, and dissolved in nitric acid, and the solution of the nitrate evaporated to dryness.

INDEX OF SEPARATIONS.

GENERAL INDEX.

PRINTED BY
SPOTTISWOODE AND CO., NEW-STREET SQUARE
LONDON

Printed in the United States
207385BV00006B/50/A